AF732621

HISTORY
OF THE
LUMBER INDUSTRY
IN THE
STATE OF NEW YORK

By

WILLIAM F. FOX

With an appendix:

THE ROLL OF PIONEER LUMBERMEN

With 22 illustrations
and a map in color

HARBOR HILL BOOKS

HARRISON, NEW YORK
1976

Library of Congress Cataloging in Publication Data

Fox, William Freeman, 1840-1909.
History of the lumber industry in the State of New York.

Originally published in the 6th Annual report of the New York Forest, Fish, and Game Commission, 1901.

1. Lumbering--New York (State)--History. I. Title.

SD538.2.N7F69 1976 338.1'7'490747 76-7561
ISBN 0-916346-23-4

Harbor Hill Books, P. O. Box 407, Harrison, N. Y. 10528

Manufactured in the United States of America

CONTENTS

ILLUSTRATIONS

History of the Lumber Industry in the State of New York.

BY WILLIAM F. FOX.

FROM the time when the pioneers first swung their axes in the primeval forests of New York, the operations of the lumbermen have been closely connected with the industrial progress and development of the State. The first settler was the first lumberman ; and his work began when he felled the trees preparatory to making the clearing in the forest where he could build his log cabin and raise his food.

Although the use of the axe alone would hardly constitute lumbering operations, as the words are understood to-day, still it was not many years until a sawmill made its appearance in each settlement, whereupon the lumber industry in all its principal details was then inaugurated.

The first colonists, owing to their limited resources and the undeveloped condition of mechanical art, went without sawmills longer than the later settlers. They obtained lumber for their houses, barns and fences by the use of the axe, supplemented at times by saws, large and small, worked by hand power. In the later settlements, however, which in the beginning of the last century included three-fourths of the State, a sawmill was built in each locality within ten to fifteen years after the first family moved in. In many instances the sawmill preceded the gristmill; and in a few places the erection of the mill antedates the advent of the first settlers.

The phrase "lumbering operations," through long usage, is now understood as including all the work of the lumbermen from the felling of the trees to the sawing of the logs. So, any history of this industry should commence properly with the erection of the first sawmills. Prior to that time trees were cut down by the settlers in clearing land and constructing their cabins. From the logs thus obtained boards and shingles were "rived out by hand," an axe and a wedge being the only tools necessary to the work. The shingles were for the roof; the boards for the floor, partitions and household furniture. This work was generally confined to the labor of one man, and to the needs of his one house and little farm. But with the erection of a sawmill the work was extended until it embraced substantially, in one form or another, all the different stages of the industry which characterize the work of the modern lumberman. Trees were felled and cut into logs of suitable length. These logs were skidded and hauled to the mill, where they were sawed into lumber of various dimensions, the product being sold to the settlers in the vicinity or floated in rafts down some stream to a greater market.

Whatever may have been the means used by the first colonists in converting trees into lumber the methods adopted in their logging operations must have been substantially the same as those of to-day.

Right here attention is called to the fact that from the days of the first settlements down to the present time the lumbermen have never resorted to clearing operations in carrying on their work of tree felling. There is, however, a widespread impression to the contrary. People are very apt to think of cleared lands and farming areas as the result of lumbering operations, and to attribute the absence of forests to the work of the lumbermen. Now, this is an error. The disappearance of the forest is due to the farmer; not to the lumberman. The farmer cuts and burns every tree and bush to make his clearing and improve his land; the lumberman takes only a few scattered trees to the acre, confining his selection to some merchantable species. Then, again, many of the fires which destroyed so much of our forest areas were due largely to the carelessness of the farmers in burning their brush and log heaps; also, but in a smaller degree, to locomotives, hunters, campers and several minor causes. The lumbermen do not start fires; for their work — chopping and log hauling — is done mostly in the late fall and winter when the woods will not take fire. Had no one ever landed on our shores but lumbermen, had no other industry but theirs ever been carried on within our borders, the forests of our State would still be standing in an unbroken, umbrageous growth.

The Primitive Forest. Its Composition.

In 1614, the year when the first houses were built at Albany and on Manhattan Island, the territory which afterwards constituted the State of New York was forest covered throughout its entire extent. Some of the Indian tribes belonging to the Six Nations had cleared small areas near their villages on which they raised corn; and in some places on the east side of the Hudson River there were openings caused by forest fires which the red men had started in order to facilitate hunting. But these clearings formed an insignificant proportion of the entire region. It was a silent, unbroken wilderness where stood a primeval forest which, in grandeur and undeveloped wealth, was unsurpassed in all the region of the Atlantic coast.

New York was not only a forest State, but it was essentially a white pine State. This valuable species was predominant throughout the entire territory. It was conspicuous everywhere by its towering height, although it may not have been as numerous as some of the inferior or smaller species with which it was associated. In height, diameter and quality of timber the pines of New York compared favorably with those of any other region on the continent. Generally they attained a height ranging from 130 to 160 feet, with a diameter, breast high, that varied from two to four feet. In some localities there were individuals of still greater size. So far as can be determined now, the maximum height was reached at about 255 feet, and the maximum diameter at about 80 inches. There is a record of a white pine which was cut in the town of Meredith that measured 247 feet in length as it lay on the ground. There are many New York lumbermen living to-day whose reminiscences include stories of giant pines that measured 7 feet or more across the stump and over 220 feet in height.

There is ample evidence as to the uniform distribution of this species throughout the State. The History of Delaware County states that "the town of Walton when first settled was heavily timbered with pine, and some hemlock, which at an early day was rafted to Philadelphia in lumber or logs, constituting the all-absorbing industry from which the land debts and living expenses were paid;" and that "the mountain east of the village of Walton received its name from the immense pines that covered its sides, and the entire valley of the village was a dense forest of the same." The History of Cattaraugus County tells of the "remarkable size and good quality" of the white pine in the southwestern part of New York. Holden's History of Warren County mentions "the splendid pines with which the great Brant Lake Tract abounded." The writer — whose grandfather and father were the pioneer lumbermen of that famous tract — well remembers hearing in his boyhood the white pine of that region enthusiastically described as being "clear as a hound's tooth." The meaning of the Indian name Schenectady — "the end of the pine plains" — indicates that this species was conspicuous in that region, although it seems there to have been largely admixed with pitch pine.

We are told that Pine Street in New York City took its name from the "many magnificent pines" that adorned the farm of Jan Jansen Damen.*

Peter Kalm,† the Swedish naturalist, who visited Albany in 1749, writes: "The white pine is found abundant here. The greater part of the merchants have extensive estates in the country, and a great deal of wood. If their estates have a little brook, they do not fail to erect a sawmill upon it for sawing boards and planks, with which commodity many yachts go during the summer to New York, having scarce any other lading than boards. They saw a vast quantity of deal from the white pine on this side of Albany, which is exported."

Michaux f. states that in 1801 "the shores of Lake Champlain appeared to be most abundantly peopled with this species."

Dr. Torrey, in his Flora of New York — published in 1843 — says: "The white pine is found in most parts of the State," and "our chief extensive forests of this noble and most valuable tree" are "on the headwaters of the Hudson, and on the rivers which empty into the St. Lawrence; on the Salmon and Black rivers, which empty into Lake Ontario; on the headwaters of the Delaware and Susquehannah; and on the headwaters of the Allegany and Genesee." This distribution includes substantially the entire State except the lowlands, from which the white pine had been taken by the early settlers long before Torrey wrote.

The Adirondack tourist of to-day can still see in the tall trees at Paul Smith's, or in the noble colonnade along the shores of Forked Lake, further evidence of its widely extended habitat.

The Catskill region was also rich in white pine, although there was a strong admixture of hemlock on its mountainous slopes. An old number of the Ulster County Gazette contains an advertisement, dated November 13, 1799, which reads as follows:

---

* New York Historic Trees. New York Times, May 12, 1901.

† The mountain laurel, *Kalmia*, so abundant in the Catskills, was discovered by this scientist and named after him.

"FOR SALE.

THE ONE HALF OF A

SAW MILL.

With a convenient place for building, lying in the town of Rochester.* By the Mill is an inexhaustible quantity of PINE WOOD.—And also,

A STOUT, HEALTHY, ACTIVE

NEGRO WENCH.

Any person inclined to purchase, may know the particulars by applying to

JOHN SCHOONMAKER, JUN."

The advertiser and his fellow lumbermen of that region have long since gone their way, and with them their "inexhaustible quantity of pine wood."

With the white pine there was in many localities an admixture of Norway pine; and in Steuben county, or along the southern border, considerable yellow pine (*P. echinata*, Mill.).

The white pine, being the most valuable of all the forest trees, was taken first, and, until 1850 or thereabouts, the work of the lumbermen was confined almost exclusively to this species.

Next, in importance, the hemlock requires some mention, as it, also, was distributed over the whole territory. Though of inferior dimensions and quality throughout the Adirondack region, yet in the southern tier of counties, and along the Catskill range, it attained a size and strength that compared favorably with the best Pennsylvania hemlock.† For a long period, however, it had no value except for its bark, which was necessary in the tanning business. Trees were cut, peeled, and the bark hauled to the tanneries while the trunks were left in the woods to decay as waste or useless material. This was largely the case within twenty-five years, especially in Pennsylvania, the demand for bark being greatly in excess of the demand for hemlock lumber. Not until there was a scarcity of pine did the lumbermen find it profitable to take hemlock logs to the mills; and, then, for many years the margin of profit was very small.

---

*Town of Rochester, Ulster County; not the city of that name.

†In the town of Colchester, Delaware County, there is a hemlock tree over two feet in diameter standing on the line between divisions 63 and 64 upon the north side of the hill opposite the school-house in the Wilson Hollow. While it was a young tree about six inches through it had been marked by some sharp instrument, probably an Indian weapon, in 1535. Two hundred and fifty-three concentric rings of annual growth over this mark is a blaze made by James Cockburn in 1788; and over another twenty-eight annual rings is a blaze made by Christopher Tappen in 1816.

The spruce, which in late years has formed so large a part of our lumber product, was confined mostly to the Adirondack plateau and the Catskill slopes. It formed no part of the forest in the western or southwestern portions of the State; neither did it appear in the southern tier of counties, along the Pennsylvania line, west of Broome county. The lumbermen who operated along the valleys of the Chemung, Tioga, Allegany and Genesee rivers cut no spruce, many of them living and dying without ever seeing a tree of this species.

It is unnecessary here to define the habitat of the other evergreen species; for the lumbermen used none of them until quite recently. The balsam, which is confined almost entirely to the Adirondack forests, is now cut to a limited extent for pulpwood and lumber; and the white cedar of that region has something of a market now in the way of shingles, posts and telegraph poles.

The hardwoods, or broadleaved trees, were present everywhere, mixed more or less abundantly with evergreens, In some places within the primeval forest there were "hardwood ridges," so called because there were no other species; in other places there were slopes on which pine, hemlock or spruce grew unmixed in pure stands or great "clumps." Along the river valleys or bottom lands where the soil was dark and rich with alluvial deposits the more valuable hardwoods — white oak, ash, cherry and black walnut — predominated. Maple, beech and birch grew everywhere, on mountain and plain; but there was no chestnut or oak on the great northern plateau, and there are none there now. Such, in short, was the general character of the forest which covered the territory of New Amsterdam when Hendrick Hudson cast anchor off the island of the Manhattoes.

Until recent years lumbermen paid no attention to the hardwoods, and but little of this kind of timber was cut, beyond what was needed for cooperage, furniture or pyroligneous acid, industries which until the present time were never prominent in this State. As fast as the lumbermen took out the pine and hemlock the great hardwood forests that remained soon fell beneath the axes of the advancing farmers and disappeared in fire and smoke.

The First Sawmills.

John Verrazzano and Hendrick Hudson made their famous discoveries and sailed away without leaving a man behind to occupy the newly found territory. No settlement was made by white men, no house erected until 1614. Just when the industrious labor of the first settlers took the form which we now call lumbering it is impossible to say. But in 1623, nine years after the first colonist

J. Y. MC CLINTOCK, PHOTO.

OLDEST SAWMILL IN NEW YORK.

BUILT IN 1805, IN ULSTER COUNTY.

built his home at New Amsterdam, three sawmills were erected there by the Dutch West India Company, and with their erection begins the history of lumbering operations in the State of New York.

The machinery of these mills, which were shipped from Holland, was constructed so that the saws could be run either by windmill or water power. One of them was built on what is now known as Governor's Island, and was probably operated by wind power; another one, which stood on Sawmill Creek, a tributary of the East River, may have used a water wheel. In 1639 the mill on Governor's Island was leased at an annual rental of 500 merchantable boards, half oak and half pine.

About this same time, perhaps a little earlier, some sawmills were built at Fort Orange (Albany) or its immediate vicinity. Andries Corstiaensen, a master mill-wright, with two sawyers, were sent there from Holland in 1630. Among the settlers at Rensselaerwyck (Troy) in 1630 were Lawrens Lawrenssen and Barent Tomassen, sawyers.* In 1636, Barent Pieterse Koeymans joined the colony, and in the fall of 1645 took charge of the Patroon's sawmills, being allowed 150 guilders a year for board, and three "stuyvers" for every plank he sawed. In two years this mill cut over 4,000 boards. In 1673 Koeymans bought a large tract of land on the Hudson River, twelve miles south of Albany—the location of the present town of Coeymans—on which there were some desirable mill sites, and where Cruyn Cornelissen and Hans Jansen had erected sawmills as early as 1651.

The colonists soon made other settlements in the Hudson Valley, and in 1661 Frans Pieters Clavers built a sawmill on the little stream which runs into the river two miles north of Stuyvesant Landing, in what is now the town of Kinderhook, Columbia County. This stream has been known as the Saw Kill ever since Clavers built his little mill there.

In 1663 a sawmill was built by Jan Barentson Wemp on the Poesten Kill, a stream which empties into the Hudson at Troy. As the falls of the Poesten Kill (*puffing or foaming creek*) furnished a strong water power it may be assumed that this mill was driven by a water wheel.

In a letter dated January 2, 1701, written by the Earl of Bellomont to the Lords of Trade, England, he says:† "They have got about 40 sawmills up in this province (the Province of New York), which I hear rids more woods or destroys more timber than all the sawmills in New Hampshire. Four saws are the most in New Hampshire that work in one mill, and here is a Dutchman lately come over who is an extraordinary artist at those mills. Mr. Livingston told me this last

---

* History of Albany County, by George R. Howell. New York: W. W. Munsell & Co., 1886.

† Colonial Documents. Vol. III, p. 825.

summer he had made him a mill that went with 12 saws. A few such mills will quickly destroy all the woods in the Province at a reasonable distance from them."

The evolution of the sawmill is largely due to the conditions and demands of the lumber industry in America. Our early colonists built and operated sawmills one hundred years or more before there was one in England. It may be well to note here that this method of manufacturing was not an absolutely necessary adjunct to the work of the forester. The wainscotings, panelled ceilings, cabinet work and Chippendale furniture which made famous the stately homes of England were constructed in all their perfection long before the first sawmill was erected in that country. The men who founded the Massachusetts Bay colony, together with the emigrants who followed them for a hundred years, had never seen a sawmill in their native land. So, if a sawmill did not always make its appearance in the colony soon after the first settlement, it does not follow that no lumbering operations were carried on. They had other means of manufacturing the forest products.

As already remarked, the pioneer of the wilderness, in building his lonely cabin, could, with axe and wedge, easily supply his few wants in this respect. But in the villages which sprang up at each important trading post there was a demand for building material and shiptimber which the inhabitants themselves could not well supply. Most of the settlers were engaged in better paying pursuits or professions; hence, some outside labor found employment in manufacturing lumber by hand power. Large timbers for house and shipbuilding were hewn out and squared with a broad-axe by men whose expert handling of this tool is now a lost art; planks, boards and boat-sides were mostly made by pit-sawing. The latter was a common industry in the old country; and one reason why England had no sawmills until after 1768 was because the mobs, who always opposed labor-saving machinery, destroyed the first ones as fast as erected, through a fear that the pit-sawyers would be thrown out of employment.

Pit-Sawing.

Pit-sawing was done by two men with a long saw which had a cross-handle at each end. A large timber, hewed square, was placed over a pit or elevated on trestles, with one man standing on top of it to pull the saw up, and one man below to pull it down. The workman on top, who guided the saw along a chalk line and was necessarily the better man, was called the "top-sawyer," a name still used occasionally in the backwoods to indicate a tip-top man; and the one below was called the pitman. When sawmills were first substituted in this work the saw was pulled up by a spring-pole overhead, and pulled down by a wooden beam attached to the

J. Y. MC CLINTOCK, PHOTO.

PRIMITIVE MILL.

IN WOODLAND VALLEY, ULSTER COUNTY, N. Y.

crank of the mill wheel. This wooden beam was called a pitman, and is known by that name to-day in every sawmill throughout the country. Our amateur philologists will not have much trouble in guessing the derivation of the word.

But pit-sawing, or "whip-sawing," as sometimes called, was not entirely abandoned on the introduction of sawmills.* This old method was still useful in sawing "long stuff," because in many mills the log carriage was not long enough to saw planks of the desired length. The writer remembers seeing two men as late as 1860 pit-sawing at the Gang Mills, near Painted Post, Steuben County, N. Y. The large, square stick of timber, which rested on high trestles, was being sawed into long planks for the sides of a canal boat. The man underneath — the human pitman — had a handkerchief tied over his head to keep the sawdust out of his eyes.

The First Lumber Shipments.

The lumbermen of New Amsterdam colony were not confined to the home market afforded by their fast growing town. They shipped part of their product to England; for, as already stated, there were at that time no sawmills in Great Britain, all of the lumber in that country being brought from Holland, or made by hand-sawyers at home. In fact, the colonists had sent some lumber to Holland in 1626, three years after the first ship load of emigrants arrived. At first, it was their only article of export besides furs. This consignment in 1626 consisted of "considerable oak timber and hickory," and was sent over in the good ship Arms of Amsterdam. In 1675 the ship Castle carried a cargo of timber valued at £400 from New York to England. In 1686 Governor Dongan, in a report to the home government, offered to "send over boards of what dimensions you please," adding that "three-inch planks for the batteries cost me fifteen shillings the hundred feet." Surely, the lumbermen of New York belong to an honorable as well as ancient guild.

For the hundred years next after the founding of the colonies at New Amsterdam and Fort Orange (Albany) the settlement of the State was confined to the region of the Hudson and Mohawk valleys. The development of the country and the growth of the lumber industry were slow as compared with that which succeeded the Revolution and the establishment of the State as a member of the new republic. There being no means of transportation except in the river districts, the lumber-

* In 1804 James Perkins built the first framed house in the town of Conquest, Cayuga County, sawing out all the lumber with a whip saw.

men, after supplying local demands, had to depend on the export business, which was confined largely to the English trade. There was a market for large white pine masts and ship timber, which gave employment to axemen and raftsmen to some extent. But four-fifths of the State was still an unbroken wilderness at the close of the Revolutionary war, and where the large and populous city of Rochester now stands there was not a house or a white man to be seen one hundred years ago.

By referring to the appendix and map published herewith, it will be seen that, aside from the Hudson and Mohawk counties, the settlement of the various districts of the State, together with the commencement of various lumbering operations, all occurred within the last one hundred and twenty-five years, while throughout a very large area nothing was attempted until a much later date. But it is interesting to notice, as at New Amsterdam, for instance, how soon the sawmill everywhere followed the erection of the first cabin, how quickly the lumber industry was established in connection with each pioneer settlement, and how closely this industry was associated with the development of the country.

The Lumber Market a Century Ago.

In 1801, according to notes made by the younger Michaux, the white pine that was cut along the shores of Lake Champlain was carried to Quebec by the Sorel and St. Lawrence rivers. He further states that "What is furnished by the southern part of the lake is sawn at Skenesborough,* transported seventy miles in the winter on sledges to Albany, and, with all the lumber of the North River, brought down in the spring to New York in sloops of 80 or 100 tons, to be after wards exported in great part to Europe, the West Indies, and the Southern States."

The same author, in writing about the lumber market at that time, says:

"The white pine is found there in the following forms: In square pieces from 12 to 25 ft. long, and of different diameters; in *scantling*, or square pieces 6 inches in diameter, for the lighter part of frames; and in boards, which are divided into *merchantable* or *common*, and into *clear* or *picked*† boards. The merchantable boards are three-fourths of an inch thick, from 10 to 15 inches wide, from 10 to 15 feet long, and frequently deformed with knots; at New York they are called Albany boards, and are sold at the same price as at Boston. The clear boards, formed from the largest stocks of the Pumpkin Pines, are of the same length and thickness as

* Whitehall.

† It may be interesting to note that "pickings" still forms one of the well-known grades made by the lumber inspectors in the Albany market.

G. H. RISON, PHOTO.

SKIDDING LOGS, HERKIMER COUNTY, N. Y.

the first, and 20, 24 and 30 inches wide. They should be perfectly *clear*, but they are admitted if they have only two knots small enough to be covered with the thumb. This wood is also formed into *clap-boards* and *shingles*."

There were foresters in those days as well as lumbermen. In 1770 Adolphus Benzel, son of Archbishop Eric Benzel, of Sweden, was appointed inspector of his majesty's woods and forests in the vicinity of Lake Champlain at a salary of £300 per annum. His residence was at Crown Point.* As early as 1700 Lord Bellomont, Governor of New York, recommended that each person who removed a tree should pay for planting "four or five young trees;" that no tree should be cut "that is marked for the use of the Navy;" and that no tree or trees be cut "but when the sap is in the root."†

Primitive Methods.

It is well to consider here the conditions and limitations under which the lumber business was carried on years ago, at the beginning of the last century, for instance.

There was a lack of the tools and labor-saving appliances which are considered indispensable to-day. Even the axe of the chopper was home made, the rude handiwork of the nearest blacksmith; for the axe factories, with their keen, brightly polished product, were yet to come. The double-bitted axe had not been invented, and so the woodsman used a single bit with a curved hickory handle. Cross-cut saws, which had to be imported from England, were scarce and costly; hence the tree trunks were cut into logs by chopping instead of sawing. The millwrights were not much better off for tools. The first mill in Rensselaer County was built in 1792 by a man named Cross, who "had no tools but an axe, saw and augur."

After the logs were cut they were seldom skidded, but were hauled direct to the mill. Skidways were rarely made, except where a surplus stock of logs was left lying in the woods.

Oxen were used for the most part in logging, the same teams being employed on farm work part of the year; for the lumberman was obliged to combine farming with his business in order to make a living. In those days the oxen did not wait for the teamster to swear at them before starting a load. It was not until later years that the oxen got that idea.

There was no river driving then. The great white pines stood close around the mill itself, and so thickly that the logs were quickly and easily "snaked" there. As

* History of the Town of Queensbury. By A. W. Holden, M. D.

† Colonial Documents. Vol. IV, p. 675.

the old-fashioned mill had only one saw, it did not require much timber to stock it, and hence several years would elapse before the haul became too long to be profitable. Then the lumberman would move his mill into another tract of timber and resume logging. It was not until years later that the Fox Brothers, the pioneer lumbermen of Warren County, conceived the plan of driving the logs to the mill instead of moving the mill to the logs, and so sent the first log drive down the Schroon River branch of the Upper Hudson.

For the first two hundred years the mills were of rude, primitive construction and of small capacity, being limited to a single upright saw. At first the saw was attached direct to the pitman, the blade being steadied by a side pressure from guide blocks. Then an improvement was made by straining the saw between stirrups in a frame or "gate," the pitman being attached to the latter. As the turbine was then unknown, the power was obtained from a single overshot water wheel.

Many of the first sawmills were built in combination with a gristmill, the power being used to drive them both or singly as needed from time to time; and in most instances both were under the same roof.

There were timber thieves in those old days as well as now. Mr. Nathan Ford, the pioneer of Ogdensburg, in a letter to Samuel Ogden, December 27, 1799, wrote: "There are several persons now cutting timber upon the two upper townships. Mr. Wilkins took down the names of several who pretended to settle; their motive was only stealing off the timber. If something is not done about this business great destruction will arise. An example ought to be made, and this cannot be done without sending an officer from Fort Stanwix. They have got the timber so boldly that they say there is no law that can be executed upon them here."

A Dangerous Life.

The life of the pioneer woodsmen or lumbermen was always beset with dangers peculiar to their work. The early town records make frequent mention of the fatal accidents which befell them. It is remarkable how often the first death in a settlement was that of some man who was killed by the falling of a tree* which he was cutting; of one who was crushed by a load of logs; or of one who was killed in his sawmill. Not only the pioneers, but their successors have contributed to the same death roll every year. The causes are various; a heavy limb falls, broken by the wind; a

*In the foot notes appended to the town histories in Hough's Gazetteer of New York there are twenty-one different instances mentioned in which the first death among the settlers was that of some man who was killed by the falling of a tree.

tree " lodges," and, springing back at the stump, kills the axeman; a load of logs "shoves" the team down some grade in the road, and the driver is thrown underneath or dashed against a roadside tree; a tier of logs starts suddenly, and a man is crushed to death; a jam on the log drive breaks without warning, and a man is lost; another, while fighting a forest fire, finds his retreat cut off and is burned; another disappears in the current of the spring flood; while in the mills there is the gruesome sight of men killed by falling on the saws.

Painful wounds and frightful accidents, also, were incidental to pioneer life in the lumber woods. In the records of the town of Middlebury, Wyoming County, we find that " In May, 1817, Artemas Shattuck went into the woods to chop. While cutting off a log that had been partially split open, his foot was caught in the crack, and he hung for a long time suspended by his foot and partly supported by one hand. Despairing of receiving aid, he finally unjointed his ankle with his pocket-knife, made a crutch of a crooked stick and started for the house."

Their privations had a pathetic side also. We read in the history of the town of Verona, Oneida County, that "the first death in the settlement was that of a child who was buried in its cradle for want of a coffin."

Rafting.

It should be remembered too that there were no canals or railroads in those days; no loading of boats or cars. The local market of each mill was limited by the distance to which the sawed lumber could be transported on wagons, over soft, newly-built roads; the greater outside market could be reached only by rafting the product and floating it down to the towns and cities which then were always located on some waterway. Hence, the mills were erected on the upper waters of the creeks or rivers, not only to utilize the water power, but to secure an outlet to their market. Each lumberman was a raftsman as well as a log-jobber and mill owner.

Passing by the lumbering operations during the first century of colonial life, of which there is very little record now, we come to a period in the history of the industry when, for a lack of canal or railway transportation, its beginning in each county was marked by the running of the first rafts.

Pioneer Raftsmen.

Arthur Noble, proprietor of the Arthurboro and Nobleboro Patents, Herkimer County, built the first mill in that county in 1790. The first lot of lumber sawed in this mill was shipped to Ireland. He rafted it down the West Canada Creek, thence down the Mohawk to the Cohoes Falls, and then carted it to the Hudson at Albany, where it was loaded in sloops for the old country.

In Broome County, 1796, Edward Edwards built a sawmill on the Onondaga stream, at a place which is now in the town of Lisle. He was the first man to run a raft down the Chenango River. For sixty years after the first settlements the staple product of this county was white pine lumber, which was rafted down the Susquehanna, sometimes to Norfolk, Va. The young men had not seen the world until they made this trip. It was a life of adventure. The river journey brought to their view whatever there was of civilization at that period; and running the dams was perilous work that furnished material for thrilling narratives on their return. Other business as well as the lumber industry was dependent on the success of the raftsmen, and notes were made payable when "the rafts got back." *

In Delaware County, Jesse Dickinson, who built a mill about 1788 on Trout Creek, in the town of Tompkins, ran the first raft that went down the west branch of the Delaware River, the lumber being floated all the way to Philadelphia.

In Chautauqua County, the first lumber floated down the Allegany River was sawed at the mill owned by Dr. Thomas R. Kennedy, on the Conewango Creek, in the town of Poland, near Jamestown. This mill was erected in 1805, and by rafting the boards in the Conewango, a tributary of the Upper Allegany, the product was taken to Pittsburg, the nearest market at that time.

In Cattaraugus County, the first lumber was rafted down the Allegany River in 1807. The rafts were owned by Bibbins Follett, Jedediah Strong and Dr. Bradley. The first sawmill in this county was built in 1801 at South Valley by the Quaker colony, and the lumber for the first raft may have been put in the river there, although in 1807 there were mills at Olean and Portville.

Every river in the State was utilized at one time or another by the lumbermen. Board rafts, bound for tidewater, or "tide," could be seen on the Chemung and Tioga rivers as late as in the sixties; and on the Upper Allegany they were a prominent feature of the lumber business until the construction of the railroad along the river shore, from Pittsburg to its headwaters in Cattaraugus county, N. Y., made this method of transportation no longer necessary. The last of them went down the river about 1880.

---

* In the town of Franklin, Delaware County, a large willow tree, which formerly stood in the highway near the house of Judge Wattles, grew from a cane which he walked home with from Philadelphia after "going down the river" upon a raft in the spring, and stuck into the ground after his return.

Rafting on the Upper Hudson.

The Hudson River was never used by raftsmen below Albany; for a raft could make no progress unless both wind and tide were favorable. So, the lumber was loaded into sailing vessels at Albany, and thus carried to New York or to the old country.

Mrs. Grant,* in describing some rafting on the Upper Hudson, as she saw it in 1768, says: "It so happened that the river had been higher than usual that spring, and, in consequence, exhibited a succession of very amusing scenes. The settlers, whose increase towards Stillwater had been for three years past incredibly great, set up sawmills on every stream, for the purpose of turning to account the fine timber which they cleared in great quantities off the new lands. The planks they drew in sledges to the side of the great river; and when the season arrived that swelled the stream to its greatest height, a whole neighborhood assembled, and made their joint stock into a large raft, which was floated down the river with a man or two on it, who with long poles were always ready to steer it clear of those islands or shallows which might impede its course. There is something serenely majestic in the easy progress of those large bodies on the full stream of this copious river. Sometimes one sees a whole family transported on this simple conveyance; the mother calmly spinning, the children sporting about her, and the father fishing at one end, and watching its safety at the same time. These rafts were taken down to Albany, and put on board vessels there, for conveyance to New York; sometimes, however, it happened that, as they proceeded very slowly, dry weather came on by the time they reached the Flats, and it became impossible to carry them further; in that case they were deposited in great triangular piles opposite our door."

The greater portion of the pine on the slopes along Lake Champlain was sent to market in rafts which were taken through the lake and down its outlet — the Sorel River — to Canada, from where it was exported to England.

In 1808, owing to international disputes, Congress laid an embargo on all trade or commercial intercourse with Canada. This restriction bore heavily on the lumbermen of Clinton County, who were dependent on the Canadian market, their lumber being sent there by way of Lake Champlain and the Sorel River. That year, a large timber raft, said to have been a quarter of a mile long, lay at Isle la Motte waiting for a favorable wind to carry it over the line. The United States revenue officer, supported by a company of militia, was watching it under orders to prevent

* Memoirs of an American Lady. By Mrs. Anne Grant. 1876. Albany: Joel Munsell.

any export of that kind. The bold raftsmen, undaunted by the display of military, pushed out into the stream, the soldiers following along the shore. After a brief exchange of musketry and rifle shots, in which no one was injured, the lumbermen crossed the line and proceeded on their way without further hindrance.

Construction of Rafts.

Rafting formed such an important branch of the lumber business for many years that some description of the methods employed seems proper here; in fact, no history of the lumber business in this State would be complete without it. In 1865, and for several years subsequent, the writer was engaged in rafting lumber on the Upper Allegany, both boards and square timber, putting in his rafts at Olean, Weston's and Portville, in Cattaraugus County, N. Y., and running them to Oil City, Pittsburg and Cincinnati. The details of the business, its varied scenes and incidents, are easily recalled to mind.

Board rafts, pine or hemlock, were from twenty-four to thirty courses deep, a "course" consisting of inch "stuff," or its equivalent in plank or joist. The number of courses varied according to the height of the "fresh," and the judgment of the pilot, on whose ability to handle the raft amid bars and shoals the owner was dependent for a successful and profitable trip. In that part of the State all logs were cut sixteen feet long, except a small proportion of twelves which were cut to save timber when necessary. Hence the platforms of a board raft were sixteen feet square. They were built up in successive layers, the boards in each course being laid at right angles to the ones just below, thus binding them together securely.

At the corners of each platform, and at intervals along the four sides, round wooden stakes, about two inches in diameter, extended from the under side of the raft to the top, passing through augur holes in some of the boards. These "grub stakes," as they were called, were cut from sapling trees, the bark peeled, and the roots cut away until only a knob remained at the lower end, a little larger than the auger hole in the boards above it. These stakes were then passed through the overlying courses and were wedged solidly at the augur holes in the binding boards of the top course so that they could not pull out. Five platforms thus constructed were connected closely, forming a "five-platform piece." The larger rafts were made up by coupling three of these five-platform pieces side by side, and fastening three more behind them. A raft of this size would then be 48 feet wide and 160 feet long; and if it had twenty-five courses it would contain over 180,000 feet of lumber, board measure.

ON THE UPPER ALLEGHENY.

RAFTS OF BOARDS TIED UP IN AN "EDDY."

The greater part of this floating mass was submerged, the top course being from six to eight inches above the water. Occasionally a "deck load" of shingles or dressed lumber was carried on top; but the latter was not taken unless the owner intended to peddle it out along the river at places where there were no planing mills.

At the forward end of each five-platform piece was a large oar. On a raft, such as just described, there were three oars in front and three behind — not on the sides, but at the ends, projecting forward and backward. The oar blade was a sixteen-foot plank, fourteen inches wide, thin on the lower edge; the oar, made from a small tree, was about eighteen feet long and eight or ten inches in diameter at the large end where it was attached to the blade. It was hewed tapering to the small end or handle at which the men stood. So large and clumsy were these rafting oars that two, and often three, men were necessary in handling them. When not in use the oars rested on the raft with their projecting blades just clear of the water. At the command of the pilot the blades were dipped by raising the handles high, and then the men, grasping the oar, with hands high lifted above their heads tramped across the platform, bracing their feet against cleats nailed on the pathway for that purpose. The men pushed, not pulled, on the oars. As the raft had no motion of its own the oars were used to move it sideways and keep it in the current; or, in turning sharp bends in the river, to swing it on its center by "crosspulling," in which the oars at one end were worked in a direction contrary to that of the oars at the other. The work of the men was directed by a pilot, whose many trips down the river had made him acquainted with every rock, reef, shoal, and bar, and whose long experience had taught him how to take advantage of the swirling current or to avoid its treacherous force.

As a large raft had six oars,—three in front and three behind,—a full crew required from twelve to eighteen men, not including the pilot and cook on one of the forward platforms. A shanty was built in which there were bunks in triple tiers around the sides. For bedding there was plenty of clean straw and coarse, warm blankets. A fireplace, on which the cooking was done, was made just outside the cabin by placing four short boards on edge and filling the square inclosure with earth.

Rafting on the Allegany.

On an ordinary flood the rafts would run from forty to fifty miles each day. At nightfall the pilot would always tie up in some eddy, swinging in to the bank, where a cable was thrown ashore and, by taking a turn or two around some large tree, the raft was "snubbed" until its momentum was slowly overcome. Yielding to the importunities of the crew, the pilot generally tied up in an eddy near some town or village, in which case there were plenty of other rafts in the same place, and the men repaired quickly to the tavern, where the question as to who was the best man and all other disputes were duly fought out. The Allegheny raftsmen were a stalwart type, many of them Indians from the Salamanca reservation; and among them were not a few "bad" men, as they are termed in modern slang. Having been paid off at Pittsburg, they generally walked home to Cattaraugus County, some of them doing easily forty miles or more a day.

It was a pleasant, jolly, outdoor life, floating down the river through the forest-covered hills and mountains of the Alleghanies, gliding past the clearings and cabins of pioneer farmers, and running through villages or cities where the bridges were lined with people waiting to see the rafts go by.

Sometimes the raft carried one or more passengers — friends of the owner or pilot — people who were content with the plain fare and food provided. The writer remembers one trip — about 1868 — when, just after passing Corydon Dam, he swung in to the bank in response to an urgent hail, whereupon a stout young fellow accompanied by a buxom girl sprang aboard. He told me that he had been married that morning, was on his "weddin' tower," and wanted to go to Franklin. He did good work on one of the forward oars, while the bride assisted in peeling potatoes and washing the tin ware.

But rafting, as well as other branches of the lumber business, had its trials and perplexities. Although the river dams were built with a chute or apron to facilitate the passage of rafts, sometimes the long floating mass would swing in the wind or current so that it would "saddle-bag" on the head of the bar below the dam. Then the boards had to be "rafted over," occasioning a delay which, on a fast-falling freshet, often resulted in getting "stuck" again on some shoal farther down the river, where the raft would lie all summer, the lower courses filled with mud and the top course warping in the sun.

Bridge piers were also a source of danger, especially where there were three or four in close succession, as at Oil City and Pittsburg. Years ago raftsmen delighted in telling the story of a Susquehanna pilot who said that there were thirty piers in the Columbia bridge, and he "run 'em all but one." The Susquehanna was a much more difficult river for rafting than the Allegany — higher dams more bridges, larger rocks and more shoal water.

It was claimed by many of the old time lumbermen that rafted lumber was better than any other, because the soaking of the boards diluted the sap and resi-

F. A. VAN SANT, PHOTO.

LOG DRIVING ON THE AUSABLE RIVER.

ESSEX COUNTY, N. Y.

nous matter so that when piled again in the yards it would season better and quicker. But whatever the advantage thus gained it was more than offset by the wet, muddy condition of the boards as they were taken out of the river. Each course had to be scrubbed with a broom; and even then the front of each pile in the lumber yard was plastered with the mud that was scraped off as the boards were drawn up over the top edge. Then, again, when the dried lumber was sent to the planing mill the boards were covered with a thin coating of dirt and grit that dulled the planer knives and filled the mill with a cloud of fine dust. And such was rafting in the old times before the railroads paralleled our rivers.

Hewed timber, as well as boards, was floated to market in rafts. Fifty years ago most of the long timber was hewed instead of sawed; for the mills had no appliances then for sawing long sticks. Moreover, the hewed timber was thought to be more valuable; it was stronger and would last longer than sawed timber wherever it was used. The sticks were of white pine, ranging from thirty to seventy feet long and from twelve to twenty-four inches square. At one time considerable "square" timber, as it was called, was sent to the Albany and New York market by canal, the rafts being made up into "lockbands" corresponding in size to the canal locks.

Log Driving.

Log driving on the Upper Hudson commenced about 1813. This method of bringing logs to a mill by floating them down a stream was first used by the Fox Brothers, Norman and Alanson, who took this means to get their timber from the Brant Lake Tract to the mills at Glens Falls, which had previously been stocked by hauling the logs direct to the mill. Their example was quickly followed, and for seventy-five years the great sawmills at Glens Falls, Sandy Hill and Fort Edward obtained their stock this way, thereby concentrating the entire manufacturing business of the Upper Hudson and its tributaries in that one locality.

In time the large number of logs coming down the Hudson on each freshet made it necessary that a sorting boom should be established at some convenient place, where the logs of the various lumber companies could be separated in accordance with the "log-marks" or characters which had been stamped or indented on the end of each log with a "marking hammer." This necessity, together with the frequent loss of stock by the breaking of poorly constructed booms in time of high water, caused the organization, in 1849, of the Hudson River Boom Association and the construction of the "big boom" at Glens Falls, where suitable arrangements were made for holding and sorting the logs belonging to all the mills located at or below that place.

From the books of this company it appears that in 1851 there were 132,500 "market"* logs, 19-inch standard, equivalent to 26,500,000 feet, B. M., received at this boom. The next year 69,080,000 ft. of logs passed through; and these figures may be assumed as showing approximately the amount of the lumber business on the Upper Hudson at that time.

The business increased yearly until the maximum was reached in 1872, when 1,069,000 markets, or 213,800,000 ft. were handled at this point. At that time the lumbermen were not cutting below twelve inches on the stump, or nothing less than "two-log" trees. Hence the logs in the boom ran, on an average, about two to the standard, and so the 1,069,000 markets delivered at the boom that year represented over two million separate logs or "pieces."

After 1872 the business declined steadily until 1900, when the books of the Boom Company showed that 282,771 markets, or 56,554,200 ft., were received that year. These figures tell briefly the story of the rise and fall of the lumber business on the Hudson watershed.

Streams Declared Navigable.

But to return to the story of the early log driving. At first, the people living along the rivers objected strenuously to the use of the streams for floating logs to the mills. The first law declaring any river in this State a "public highway" was passed in 1806. This act provided that the Salmon River in Franklin County could be used for rafts and boats below Malone, and it enacted further that if any person shall "cut or fell any trees into the said river such person shall forfeit one dollar for each tree so felled and suffered to remain in the said river twenty-four hours." This same law — chapter 139 — forbids any person from "rolling any log or logs into the Schroon River, in Essex County, or doing anything to obstruct said river," under a penalty of five dollars for each offense; but provides that nothing in the act shall "prevent any person from rafting any lumber down said river they may think proper." The restrictions in this law as to obstructions will be read with interest by those who in recent years have noticed how often some Adirondack river is filled at times for a long distance with a solid mass of floating logs through which no boat can pass.

---

* The market or standard log is 19 inches in diameter at the small end and 13 feet long, and is generally referred to as a "market" by the lumbermen of the Eastern Adirondacks.

LOGGING IN WINTER.

But industrial interests are always recognized in time, and so, in 1854, the Legislature delared the Salmon River "a public highway for the purpose of floating saw-logs and timber." In fact a portion of the river was so used before the passage of this law.

In 1804 the Legislature passed a law, chapter 103, to punish any one who stole timber or lumber that was floating down the river or lying along the shore. This act refers to "any timber, hewed, sawed or riven," terms which do not seem to include saw-logs, and which would indicate that only long timber, spars and masts were floated down the stream at that time. Section 2 of this law provides a severe penalty for persons who shall "deface or alter any mark, or put a false mark on any such timber," from which it appears that "log marks" were in use then, even if short logs were not driven down the stream. Nine years later a law was passed — chapter 34, Laws of 1813 — requiring all log marks to be recorded in the office of the town clerk of Queensbury, and the phraseology of this act shows that log driving had commenced already on the Hudson and "its branches to the northwest of Baker's Falls" (Sandy Hill.) In 1825 a similar act was passed for the protection of log marks on the Au Sable River.

In 1810 the Legislature declared the Raquette River a public highway from its mouth to the falls; but the text of the act specifies "rafts or boats" only. The same law opens up a portion of the St. Regis River, and provides that dams shall be constructed so as to permit the passage of "timber, masts and spars." In 1850, chapter 249 provided for the passage of saw-logs and appropriated money for "clearing and improving" the channel.

The Black River was first declared a highway in 1821, the law providing for "rafting timber or lumber." The Grasse River was made a highway in 1824; and in 1854 an act was passed "to improve" this stream "for floating timber and saw logs."

The years in which other rivers were declared public highways for floating logs are as follows:

Genesee River, *	1818
Delaware River,	1823
Saranac River,	1846
Moose River,	1851
Chateaugay River,	1851
Beaver River,	1853

* In 1828 the Genesee River was declared a public highway from Rochester to the Pennsylvania line.

St. Regis — West Branch,	1854
" East Branch,	1860
West Canada Creek,	1854
Oswegatchie River,	1854
Sacandaga River,	1854
Great Chazy River,	1857
Deer River,	1867

After the merchantable timber along the main rivers had been cut, the lumbermen turned their attention to the more remote and inaccessible tracts on mountain slopes, where the streams were narrow, rocky and rapid. Then commenced the erection of "splash" or "flooding" dams, which were used to drive the logs out of the small streams, on the temporary, artificial floods caused by opening the gates; and, also, to reinforce the subsiding waters of the main streams.

These flooding dams seldom did any damage to standing timber, because the ponds were always drawn down in the early spring when water was needed for log driving, the gates being left open until the next winter. There was no backflow during the period of vegetation; and the temporary flooding of the roots of trees does not kill the timber. Trees are killed by water only where it is allowed to cover the ground for two or more successive summers.

There is a general impression, however, to the contrary; and it is commonly believed that the lumbermen with their flooding dams are responsible for the killing of live timber and the destruction of forest scenery. But, in nearly every instance, the dead timber in the flowed lands of the Adirondacks is the result of some dam or reservoir which was built in the interest of State canals, local steamboat lines or manufactories on the lower waters. The lumbermen had little or nothing to do with them.

In the southern and western portions of New York the lumbermen rarely built these small flooding dams to assist them in their river driving. The country was not so mountainous; the streams were not so rapid or violent as in the Adirondack forests; the spring floods held up longer by reason of a less rapid flow; the log driving was easier in every respect.

Log Drivers and Their Work.

The beginning of log driving was coincident with the sudden increase in the development of the country at the commencement of the last century. As there were no canals or railroads, and as logs could no longer be supplied under the former primitive methods of hauling from the forest to the mill, some such method became necessary owing to the corresponding demands for lumber. The haul had become

F. A. VAN SANT, PHOTO.

CAMP OF ADIRONDACK RIVER DRIVERS.

TRAVELLING KITCHEN, OR COOK-SHANTY IN BACKGROUND.

too long, forcing the manufacturers either to move their sawmills up stream or to flood the logs down to some place where they could be sawed.

In a few years log drivers were at work on every large river in the State. The logs, which were cut and skidded in the fall, were hauled during the winter to the banks of the streams, where they were piled in huge tiers on the "banking ground," as it was called on the Susquehanna, or "rolling bank" in Northern New York. With the first spring freshet, often while the ice was still running, the blocking was knocked loose, allowing the great piles of logs to roll down the sloping ground, over the edge of the bank, in a swiftly moving, tumbling, splashing mass, into the cold, turbid stream. In some places the logs were unloaded from the sleighs directly on the ice which covered some lake, pond or stream, with the intention of allowing the logs to go out with the ice on the first spring flood.

In the lake region of the Adirondacks, on the Saranacs and neighboring waters, the river drivers had the additional task of moving their logs through the lakes, where there was no current to assist their progress, and, too often, contrary winds that drove their logs back or scattered them. In passing through these lakes the lumbermen generally rafted the logs or inclosed the floating mass within strongly connected booms, and then "warped" their way through the open water by using an anchor, a long heavy cable and an upright windlass which stood on the forward end of the raft. This work was often done at night when the lake was still and free from the strong winds that are so prevalent in early spring. There are old river drivers living to-day who, in their stories of early log drives on Floodwood and the Upper Saranac, describe how, through the long hours and darkness, they leaned wearily against the capstan bars as they tramped round and round the platform while "kedging" their way down the lake.

The work of the river drivers was a perilous one. Scarcely a season passed without some one being drowned or killed on some stream. Men were crushed under swift rolling logs at the banking grounds; chilled to death in the icy waters; drowned while working at the rapids; or killed in breaking the great jams which formed at every obstruction in the river. The most dangerous work was often done by volunteers, and could all the tales of the river drivers be collected there would be found stories of unrecorded heroism that would equal anything in fire and flood and battlefield. They were necessarily men of stalwart build and superb physique. Possessed of a surprising agility, they would leap from log to log in the rapid, swirling current, and with nothing but a pike pole for a paddle would ride a small log, standing upright and guiding their treacherous craft as skillfully as an Indian his canoe.

But the river driver is passing. Other methods of getting logs to the mill are fast coming into use. And now when the hunter or fisherman makes temporary quarters in some old abandoned logging camp he looks thoughtfully at the puncheon floor, thickly pitted with the marks made by the spikes in the river drivers' shoes, and thinks of the old days of the big log drives, or listens to the guide as he tells the story of some "good man" who met death bravely in the foaming rapids or on the head of the jam.

On the larger streams the owners of the various mills generally arranged for a "union drive," in which expenses were shared according to the number of logs each one had in the drive, a proportion which was easily ascertained from the tally at the sorting boom or from the books of the log scalers.

Log Marks.

Each lumberman on the river had his own peculiar log mark, which was stamped with a marking hammer on the ends of his logs while piled in the woods on skidways, or before they were put into the stream. There were so many different firms operating on the Upper Hudson that the ingenuity of the lumbermen was greatly taxed to devise new and distinctive characters to mark their floating property.

Some of the principal marks used, from 1851 to 1900, on the Upper Hudson and its tributaries — the Sacandaga, Schroon and Boreas rivers — were as follows:

A. Wing & Co.	⊕
A. N. Cheney	
Tefft & Russell	○
Morgan Lumber Company	♡
James Morgan & Co.	
Finch, Pruyn & Co.	£
Morgan & McEchron	OK
W. H. Bloomingdale	X
D. W. Sherman	S
A. Sherman	B
George H. Freeman	□
Bradley & Underwood	DU

J. Y. McCLINTOCK, PHOTO.

RAQUETTE POND, ON RAQUETTE RIVER.

NEAR THE VILLAGE OF TUPPER LAKE, FRANKLIN CO., N. Y.

Kenyon Lumber Company 

Orson Richards 

Van Dusen & Crandall 

Freeman & Van Dusen 

Cheney & Armes 

Lemon Thomson 

Thomson, Douglas & Dix 

Union Bag and Paper Company 

International Paper Company 

On the Tioga River, the following marks were used to distinguish the logs coming from various timber tracts:

Fox, Weston & Bronson 

Ballard & Co. 

Cameron & Co. 

On the Oswegatchie River the logs of the principal firms were stamped as follows:

Starbuck & McCarty 

Weston, Dean & Aldrich 

On the Indian River (Lewis Co.), Roberts & Brooks used this mark: 

On the Raquette River the log marks recorded in the town clerk's office at Potsdam since 1851,* included, among others, the following:

Ransom Jenkins 

A. M. Adsit & Co. 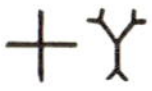

Henry Hewett 

Hewett & Townsend 

Parmenter & Hitchings 

*The law requiring that the log marks on the Raquette River should be recorded was passed in 1851. It allowed the town clerk twenty-five cents for recording each mark. Many of these marks had been in use on the Raquette River prior to 1851. Since then 102 different marks were recorded, the last entry having been made December 4, 1900, by the Raquette River Paper Company.

George L. Stanton	
Morgan, Rosekrans & Adsit	
S. Chamberlain & Co.	
Hitchings & Hall	
N. Pratt & Son	
Ralph Pomeroy	
Archibald Robertson	
Ralph & Co.	
George W. Sisson	
Sherman Lumber Company	
Augustus Sherman	
Burnham, Loveless & Co.	
M. S. Ballard	
W. A. Griswold	
Foster & Boswell	
Harmon & Rice	8
Norwood Manufacturing Company	
Watkins, Turner & Co.	
Export Lumber Company	SAX
Export Lumber Company	XPT

In addition to the foregoing there were a large number of log marks on the Raquette in which letters or numerals were used.

On the Saranac River there were:

Maine Company	XX
Christopher F. Norton	
H. & O. A. Tefft	
J. H. & E. C. Baker	5
Everitt C. Baker	

STEAM LOG LOADER.

(Courtesy American Hoist & Derrick Co., St. Paul, Minn.)

Loren Ellis	E
Thomas & Hammond	T
Patrick Hanlon	OO
D. H. & W. Parsons	

As these firms went out of business from time to time, their marks were used by other lumbermen, the original owners having no further use for them.

To reprint here all the marks used on the many logging streams of New York would occupy too much space. The ones shown here will give some idea of what constitutes a good log mark. The characters were usually about an inch and a quarter in length or width. In some instances numerals were used. Jones Ordway, one of the pioneer lumbermen of Northern New York, stamped his logs with a 34; and Harris, Finch & Co. used a 15.

The river drivers and men at the sorting booms had their own names for the various log marks, designations which were suggested by some real or fancied resemblance. Hence, in speaking of log marks we hear the river drivers use the names crow's foot, double O, wine cup, triangle, hawk eye, devil's head, deer's foot, etc.

In making the design for a mark care was taken to select one that would be quickly recognized as the log passed through the opening at the sorting boom; and the man who used the marking hammer always hit the end of a log in several places so that whatever side might be up as it floated, some one of the marks would be in sight. These marks were also of use in the mill yard or piling ground, as showing the job or firm to which a pile of lumber belonged, a part or all of the stamp showing here and there on the ends of a board or plank.

Log Railroads.

But with railroads, logging cars, steam log loaders and jack-works for lifting logs from a lake or stream onto the cars, lumbermen are no longer dependent on river driving in order to get their stock to the mill. A railroad has the great advantage of being available any month of the year. The mill does not have to shut down in a dry season for lack of stock. There is no loss by having a drive "hung up" for many months, the logs decaying or deteriorating in the meantime. With the railroad the stock is brought to the mill in perfect condition — no faster than wanted, but just as fast as it is wanted.

The first railroad for hauling logs was built in 1860 by Fox, Weston & Bronson, in the town of Lindley, Steuben county, N. Y. It was constructed of wooden rails, and was equipped with platform cars and a locomotive which bore on its cab the name " Bull of the Woods." This railroad was not used as a substitute for log driving, but for hauling logs to the bank of the Tioga River, where they were rolled into the stream and driven to the large gang mills at Painted Post.

Railroads were utilized more and more, improvements were made in the mechanical appliances for loading logs on cars, and in time it became evident that railroading was more economical in the end than river driving. Our Adirondack lumbermen have been quick to perceive the many advantages of this new system of bringing stock to their mills, and now there are several timber tracts in Northern New York from which all the logs are taken out by rail. A central line is built through the forest, with branch tracks and switches in place of log roads. The tourist on our Adirondack trunk lines may see any day long trains of flat cars loaded high with logs bound for some sawmill or pulpmill.

The use of railroads in the transportation of logs enables the lumbermen to get out their hardwood timber, which otherwise would be inaccessible and worthless because it cannot be floated down the streams. The logs are too heavy; they would sink. The hardwoods — birch, maple, beech, ash and cherry — constitute on an average over sixty-five per cent of our northern forests. In the Catskills these species form a still larger proportion. As this class of timber becomes accessible under improved methods of logging, the value of such timber lands is greatly enhanced.

The increasing use of railroads by our lumbermen and consequent rise in the price of hardwood lands will have a direct influence on the forest policy of the State. Hitherto the State has been purchasing forest lands at a low price, because the lumbermen, having removed the small proportion of conifers — spruce, pine and hemlock — were willing to sell at a low figure. The remaining hardwoods could not be marketed, and the taxes were burdensome. But now that the hardwoods, as well as the others, will probably be cut, there arises an imperative necessity for more prompt and liberal action on the part of the State Legislature if our forests are to be preserved. Hitherto lumbering has meant a culling or thinning process, the removal of the evergreens only; it soon may mean denudation.

Length of Logs.

The lumbermen of Southern and Western New York cut their logs mostly into lengths of sixteen feet to supply the market demand for sixteen-foot boards. They cut some twelve-foot logs in order to save timber, and there was generally

G. H. RISON, PHOTO.

MEASURING AND MARKING LOGS.

enough of these short logs to supply the call for boards of that length. On some jobs fourteens were cut to save timber; but boards of this length were somewhat unsalable.

In the Adirondack forests all logs, with scarcely an exception, are cut thirteen feet long. The reason for adopting this odd length is not known now. For nearly a hundred years the lumbermen of Northern New York have cut thirteen-foot logs, sawed thirteen-foot boards, and sold thirteen-foot lumber in the Albany and New York markets, although logs everywhere else in the United States are cut into lengths of sixteen feet or some other even number. Fifty years ago the ten-inch boards, thirteen feet long, from the Glens Falls mills, were known in these markets as "tally-boards," and were sold by count instead of measure. It may be that, originally, the liberal-minded lumbermen of that region considered their logs as twelve feet long with an extra foot added for good measure. However this may be, the boards are now measured at their full contents.

Log Rules.

Throughout the entire State, with the exception of the Adirondack country, the lumbermen bought or sold logs by one of two rules — Doyle's or Scribner's. Many years ago, prior to 1850, Edward Doyle and J. M. Scribner each published an original tabulation of figures — called a log rule — showing, according to the careful calculation of the author, the number of feet, board measure, which a log of any given size would yield when sawed into inch lumber. For half a century or more the relative merits of these two log rules have been a source of frequent discussion, there being a material difference in the figures given by these authorities.

It may seem strange that there need be any dispute over a mathematical problem of this kind. Certainly if saws were of the same thickness and sawyers equally skillful there could be no variation in the results. But logs are not cylindrical; they are tapering, sometimes crooked, often rotten in spots, and apt to be defective in various ways. Hence, in formulating a log rule for general use a question arises as to the proper allowance for slabs, sawkerf, waste and the various defects that may be found in almost any large lots of logs. It was on this matter of allowance that Doyle and Scribner differed.

The Doyle rule is based on a fixed, arbitrary formula which is fairly correct as to medium-sized logs, but is inaccurate, necessarily so, as to others. The formula is this: Subtract four from the diameter in inches; square one-fourth of the remainder, and multiply the product by the length in feet. Hence, for a 24 inch log 12 feet long the process is as follows: $24 - 4 = 20$; $20 \div 4 = 5$; $5 \times 5 = 25$; 25×12 (length in feet) $= 300$, the number of feet in the log. The same result can be obtained by another method. Starting with the sixteen-feet lengths, the number of feet, board measure, in a log of any diameter, is found by deducting four inches from the diameter and then taking the square of the difference. For instance: in finding the contents of a log sixteen feet long, twenty inches in diameter, use this formula, $(20'' - 4'') \times 16 = 256$ ft. Having determined the contents of the sixteen-feet logs, all other lengths are computed proportionately. Hence, a log twelve feet long and twenty inches in diameter would contain 192 feet, or three-fourths of 256.

Scribner arrived at his table of contents by platting circular diagrams showing the number of square edged boards in a log properly sawed.* From these diagrams the amount of lumber was computed for each diameter, after which a fixed percentage was deducted for the imperfections which are common to an ordinary lot of logs.

Doyle's figures for the contents of small logs are much below those given by Scribner; while in large logs he allows more than Scribner. Both rules agree substantially on logs of 24 inches in diameter, the lines crossing at this point and diverging in opposite directions. For instance:

DIAMETER.	LENGTH.	DOYLE.	SCRIBNER.
10 inches.	13 feet.	29	41
24 "	13 "	325	328
36 "	13 "	832	750
10 "	16 "	36	54
24 "	16 "	400	404
36 "	16 "	1,024	923

As the greater part of the trees, and all the top logs, were less than twenty-four inches in diameter, the sawmill owners, being the purchasers, naturally favored the Doyle rule. The logs were bought or contracted for with jobbers, at a fixed price per thousand feet; and under the Doyle rule the stock cost less money. In letting log contracts to their jobbers, or in purchasing from outside parties, the sawmill companies insisted on buying, contracting, and measuring by the Doyle rule. As a result, the old Scribner rule was gradually abandoned, and is rarely used at present. In fact, Scribner himself discarded it adopting the other and publishing it under his name as the new Scribner or Doyle-Scribner rule. The publishers claim that over one million copies of this book have been sold, it being in general use by lumbermen throughout the entire continent.†

On a lot of straight, sound white pine logs, ten to twenty inches in diameter, Doyle's rule would be too severe; the contents as shown by the "saw-bill" would

*See diagram, p. 70, Scribner's Lumber and Log Book. Rochester, N. Y. Geo. W. Fisher, 1893.

†Published by George W. Fisher, Rochester, N. Y., 1893.

overrun the log measurement. But with a lot of hemlock logs which, when sawed, run largely into culls and waste because of shaky timber and other defects, the Doyle rule is necessary if the purchaser expects his saw-bill to hold out, and to have a million feet of lumber for each million feet of logs bought, or delivered by his jobbers. The writer has had several opportunities to test this matter on a large scale.

In the Adirondack forests there is no log rule in use which gives the contents of a log, like the Doyle or Scribner. Throughout this entire region logs are bought and sold by count, a certain size called the standard being adopted as the unit. Logs of a smaller size are counted as a proportionate fraction, and the larger logs as one and a fraction, two and a fraction, or more. The standard of count in general use is a log 13 feet long and 19 inches in diameter at the top or small end. The logs are all measured, and in counting them each log is reckoned according to the ratio which the square of its diameter bears to the square of the diameter of the standard or nineteen-inch log. This is obtained by dividing the square of the diameter of each log by the square of the diameter of the standard, the divisor in each case being 361, which is the square of 19.

For instance: the square of 16 is 256, which divided by 361, the square of the standard, gives .709. Hence, a sixteen-inch log is counted as .709 of a standard. A nineteen-inch log counts, of course, as one. A twenty-eight inch log counts as 2.17 standards ($784 \div 361 = 2.17$).

This is what is called the Standard Rule, although it is not a log rule in the same sense as Doyle's or Scribner's; for it does not give the contents of logs, but is merely a method of counting that can be done by anyone. Printed tabulations showing the fractional amount of each diameter are now largely in use, and the log scaler no longer works at night in the "men's room" of the shanty extending the figures in his tally-book by the light of a lantern.

If one wants to know the contents of a log, board measure, he must still turn to some log rule. Doyle gives the contents (board measure) of a 19-inch log, 13 feet long, as 183 feet; while the old Scribner rule allows 195 feet. A sound, straight log of this size, under a band saw and a good sawyer, will yield 200 feet of straight edged boards.

On the Saranac River, also, the lumbermen buy and sell their logs by count, but use a 22-inch diameter as a standard. This is known as the Saranac standard; the 19-inch log is called the Glens Falls standard. Lumbermen often use the word market in place of standard, and speak of "market" logs; of twenty thousand "markets" in a drive; or of letting a log job of fifty thousand "markets."

Adirondack lumbermen always estimate five markets as equal to one thousand feet of logs; and so a log job, or drive, of fifty thousand markets may be accepted as equivalent to ten million feet, figures which are more easily understood by an outside lumberman.

The idea of buying and selling logs by count, using some fixed size as a standard unit, originated with Norman Fox, a pioneer lumberman of Warren County. For several years he represented that district in the State Legislature, where, by reason of natural ability and a collegiate education, he attained prominence in council and debate. Subsequently, in his lumber operations in Upper Canada, he adopted a twenty-inch standard, the square of which makes a more convenient divisor.

The methods or rules of measurement were never made the subject of any legislation, neither for logs or sawed lumber. There were laws providing for the appointment of inspectors and defining the number that might be appointed in each of the various districts into which the State was divided for this purpose. In 1805 an act was passed for the inspection of lumber, rafts, timber and spars, which allowed the inspectors to charge 37½ cents per 1,000 ft. B. M., and 14 cents per forty cubic feet. The inspectors were required by this law to mark all lumber or timber, which they had inspected, with a "marking iron," showing the number of feet in each piece.

Sawmills.

As already described, the first sawmills in each locality throughout the State or Colony were of primitive character, containing one upright saw for which the power was furnished by an overshot water wheel. In time it followed naturally that an additional saw was inserted in the gate, and so on until the modern gang was evolved.

The first gang mill was built on the Hudson River at Fort Edward; just when, the records do not show. It is claimed that the next one was erected in 1848 by Hinckley & Ballou, on the West Canada Creek, in the town of Russia, Herkimer County. But prior to this year there were gangs running in the mills at Glens Falls, Sandy Hill, and Painted Post.

In 1848 Henry S. Shedd and Marshall Shedd, Jr., erected a gang mill in Lewis County at the lower falls of the Moose River, about one mile from its junction with the Black River. The gang in this mill contained thirty-two saws.

The first gang mill on the Raquette River was built at Norwood, in 1851, by Morgan, Rosekrans & Adsit.

For many years after their introduction gang sawmills were operated by water power exclusively. Their owners, having secured the best mill sites on our rivers,

J. Y. McCLINTOCK, PHOTO.

A MODERN GANG. NORWOOD MFG. CO., TUPPER LAKE, N. Y.

THIS MILL HAS TWO GANGS, TWO BANDSAWS, AND TWO CIRCULARS.

did not need to run their mills by steam. Water power was cheaper; and many of the largest gang mills in New York use it to-day.

The large mills have not changed materially in the last forty years. In 1860 there were several in this State that ran five or six gates; say, three gangs, a slabbing gang, and two English mills.* Some used a Yankee gang instead of a slabber; and in the large mills, where two or more stock gangs were in use, one of them would contain about thirty-six saws, set for inch boards, while another would be hung with a smaller number, set for plank. Such a mill — six gates — would cut about fifteen million feet per year, running night and day, with an occasional shut down for low water.

Steam mills as a class used a large circular saw, although there are now mills in New York, as elsewhere, which run both circulars and gangs, and of late years band saws, also. It is difficult to find any records showing when the first steam sawmill was built in this State. A sawmill driven by steam power was built in 1830, in the town of Newark Valley, Tioga County, by Chester Patterson and Jonathan Day, which employed about thirty men. The engine had a walking beam, such as is used on steamboats.

In 1833 George Kirby erected a steam sawmill in the town of Nichols, Tioga County. In 1842 Dexter and Daniel Davis built one in the town of Caton, Steuben County; in 1844 a steam sawmill was erected in the town of Hammond, St. Lawrence County, by James E. Lyon; the first steam mill in Erie County was built at Tonawanda, in 1847, by Col. L. S. Payne; and in 1849 one was started by Kitts & Broadway in the town of Denmark, Lewis County.

Tanneries.

Although the lumberman has little in common with the tanner, yet the demands of the latter had a material influence on that part of the lumber industry which included the hemlock trade. To-day hemlock lumber finds a ready market, and at a price equal to that paid for spruce not many years ago. But a large portion of the hemlock in this State was cut by lumbermen to supply the demand for bark only, the logs being left in the woods to decay and waste. This was particularly the case in the Catskill counties, where this species was more abundant than elsewhere in the State; only the most accessible of the timber was hauled to the mills.

Bark peeling in the Catskills ceased prior to 1870, and the great tanneries at Phœnicia, Woodland, Shandaken, Big Indian, and Prattsville had to abandon their

*The English mill is an ordinary square gate or frame containing one or two upright saws, with a sixteen-foot carriage that gigs back.

business for lack of bark. The tannery at Prattsville, owned by Zadoc Pratt, was one of the largest in the State. Here, 60,000 sides of sole leather were tanned and 6,000 cords of bark consumed annually for twenty-five years. Over six million feet of hemlock was cut each year to furnish bark for this tannery alone, the greater part of the logs being left to rot after they were peeled.

The principal men in the tannery business in the Catskill region were: Col. H. D. Snyder, Phœnicia; James Simpson, Phœnicia; Pratt & Sampson, Shandaken, and Zadoc Pratt, Prattsville. In 1865, according to the State census, there were 820 tanneries in New York; to-day there are not a dozen all told.

Methods of Lumbering.

The system employed at present in logging and lumbering is substantially the same as that used by the pioneers; it is based on the methods which evolved from the trials, failures, successes, and general experience incidental to the early years of this industry. Of course, improvements have been made in tools and mechanical appliances; and the men are better housed, fed and paid. But the general principles on which the business is conducted to-day remain the same, and so some information as to the details of the work at the present time may give some idea of how it was carried on a century or more ago.*

When the lumberman, landowner, or sawmill man, as the case may be, decides to lumber some certain tract, he lets a contract to cut the logs of a particular species and deliver them on the banks of the stream or lake, whence they are to be floated to the mill. This is called letting a log job, and the man who contracts to do the work is called a jobber. In some places the contract is let by the thousand feet; but in the Adirondacks it generally calls for so many thousand standards or markets. The agreement is written out in duplicate and signed by both parties. Having let the contract, the principal generally goes to some bank, where, by getting his notes discounted, provision is made for meeting the expenses of the business and making the advances in cash which the jobber is very apt to ask for.

The jobber commences operations by the erection of his logging camps, which are located on the tract and as near as possible to the timber which is to be cut. The buildings or " camps " are made of logs, the cracks being well " chinked " with

*A good description of life in the lumber camps as it existed many years ago may be found in the interesting volume, Forest Life and Forest Trees, by John S. Springer. 1856. New York, Harper & Brothers.

G. H. RISON, PHOTO.

AN ADIRONDACK LUMBER CAMP.

sticks and moss to keep out the wind and cold. A "tote" road in the meanwhile is cut through the woods, over which are hauled loads of boards for use in the construction of buildings, and the making of the necessary partitions, doors, floors, bunks, and the long tables at which the men eat. In old times no floors were laid, the earth being leveled off to serve the purpose; but now most of the camps have floors made of boards or flattened logs. Shingles and window sashes are also hauled in, and the camps are made ready for their occupants. On the larger jobs they are built to accommodate from eighty to one hundred men.

There is generally one large low building, with an attic fitted up with tiers of bunks for a sleeping room, the ground floor containing a large room filled with long rough tables on which the meals are served. The front end of this room is partitioned off for a "men's room," where the crew sit evenings, smoking, reading, singing or card playing, preparatory to climbing the ladder to their night's rest in the attic. Of late years women are often employed in the lumber camps as cooks. Hence the name "men's room;" for the crew are not allowed in the eating room or kitchen except at meal time. A violation of this rule is apt to evoke the displeasure of the cooks, and a dipperful of hot water as a penalty for "snoopin'" around the kitchen and talking to the women folks. In most of the Adirondack camps, however, a French Canadian officiates as cook, with another man for an assistant who is known in some camps as the "cookee."

Another log building, one story only, is built for a barn, in which the numerous teams of horses employed on the job are warmly and comfortably cared for, and where a stock of hay and oats is stored for their maintenance. The blacksmith being an indispensable man in a logging camp, a log building of suitable size is put up for his use, in which there is a forge and all the tools for shoeing horses, mending chains and repairing sleds. In the larger camps, where a big job is being carried on, an additional building, known as the office, is erected for the convenience of the boss, the log scaler, and the timekeeper, who have their rude desks there at which they work evenings. Sometimes, also, there is kept on hand a small stock of necessary articles for sale to the men, such as shoes, stockings, mittens, tobacco, etc.

The "tote" road having been put in a fairly passable condition, the jobber hauls in his stock of provisions, tools, and feed for his teams, followed by straggling groups of hardy looking men, most of whom, having spent their previous earnings in some metropolis of the wilderness, have no alternative from another long sojourn in the lumber camps.

If the contract of the jobber includes hemlock and bark peeling, work opens in the early summer, for the bark will peel only from May 20th to August 20th, or thereabouts. For this work the crews are divided, each man having his particular task assigned him. The best axemen are detailed for the felling of the large hemlocks. Others with their axes girdle the fallen tree trunks at intervals of four feet, and these are followed by men with "spuds," an iron tool with which they peel or pry loose the bark. The first "ring," the one at the base of the tree, is taken off before the tree is felled; otherwise, the cutting at the stump would spoil this piece of bark. Another gang works as "swampers," or in piling and ranking the bark ready for hauling.

With the approach of autumn the sap ceases to flow; the bark consequently sticks to the tree, and the work of peeling is ended for that year. The axemen now turn their attention to the spruce, pine and balsam, and the forest echoes with the constant crashing of falling trees. As before, the fastest choppers are detailed for cutting down the trees; they set the pace for the whole job. These men show wonderful expertness in cutting a stump so that the tree will fall exactly where it is wanted, some of them being able to set up a stake as a mark and drive it into the ground with the falling tree trunks. Some such skill is necessary to the work in order to prevent a tree from "lodging" in another one as it falls, to avoid piling it on other down timber, and to protect as far as practicable any young growth of saplings which may be standing near.

While the axemen are busy dropping the trees others are employed in lopping off the limbs from the fallen trunk, and these are followed by men with long crosscut saws who divide it into logs of the specified length. Then come the men with teams who haul the logs to the skidways and pile them up there in huge tiers, ready to be loaded on sleighs when snow comes. The skidways were formerly made of spruce or balsam and left in the woods to decay. But now, owing to improved methods, and the increased value of the timber, hardwood skids are in common use. If it becomes necessary to use softwood skids they are cut up into logs and taken out at the end of the job. While on the skidways the logs are measured, or scaled as it is termed, the scaler taking the diameter of each log, within the bark at the top or smallest end, and jotting it down in his book, after which a man with a marking hammer stamps the logs on both ends with the owner's mark. The scaler generally has an assistant; for the logs in the large piles must be measured at both ends in order to determine which is the top, the body of the log being out of sight. Crooked or rotten logs are "docked" in measurement enough to offset the defects.

While the cutting and skidding is going on, the jobber, aided by another crew, is busy with the work of building roads leading from the skidways to the main

G. H. RISON, PHOTO.

SAWING FALLEN TREES INTO LOGS.

road, which in turn leads to the lake or river bank where the logs are to be unloaded. A diagram of the log roads on a big lumber job would resemble a tree with subdividing branches, although a somewhat crooked one owing to the curves and windings of the ravines or depressions down which the roads must go. This laying out of the roads is an important part of the work, for upon the skill and judgment exercised in it much of the profit in the job depends. All necessary roads must be built, but unnecessary ones must be avoided. The jobber must exercise no little engineering skill in selecting a line that will reach all his skidways, and at the same time preserve a practicable grade. It must be down hill all the way from the starting point, so that large loads can be hauled, and yet not so steep as to shove a team over the bank. Skill and experience are also called for in the construction of side hill or dugway roads, in bridge building, and in corduroying swamps.

The camps have been built, the bark peeled and ranked, the skidways piled high with logs, and on the first deep snow the hauling commences. The roads, having been scraped smooth, are sprinkled each night from a large water tank, drawn on a sleigh, until a good ice bottom is formed. Brush, straw, or sawdust has been strewn on the steep grades to retard the speed of the loaded sleighs, and the banking or rolling ground soon becomes a scene of activity as the teams drive up in quick succession to be unloaded. The binding chains are quickly unfastened by ready hands, and the huge pile rolls off the sleigh with a bumping, thumping noise as the logs rebound from the frozen earth.

The teamsters vie with each other in the size of their loads; and, with the wide "bunks" now in use, iced roads, and heavy teams, a pyramid of logs is rolled up on the sleigh at the skidways until the driver, astride of the top log, is perched ten feet or more above the ground. Some of the "champion" loads contain from five to six thousand feet, although much smaller ones are the general rule. A teamster is expected to make a specified number of trips each day, according to the length of the haul, which varies in most jobs from one to four miles. He needs go to sleep early, for he must be up long before daylight, feed his team, get his breakfast, and be off on the road while the stars are still shining clear in the cold winter sky.

The wages paid in the lumber camps of Northern New York run about $28 per month, including board, with a distinction in favor of first class men, to whom higher wages are paid. The jobbers are quick to recognize a good hand, and a man is paid what he can earn.

With the delivery of the stipulated number of logs at the river bank the jobber has finished his contract, and he "goes to town" to settle up with his employer. The men have been paid off and have taken their way to their homes or to some village, where too many of them soon part with their hard earned wages in whiskey drinking and riotous scenes. Old grudges are fought out, and, not infrequently, some luckless hero of the camp spends his vacation in the county jail. As a class, however, they are honest, brave, and industrious, reflecting credit on the great industry with which their life and labors are so closely identified.

But the winter is soon past, and the spring thaw with its warm south winds and rains is rapidly loosening the ice in the upland streams and lakes. The boss river driver collects his crew of stalwart, daring men, and they again betake themselves to the woods, where the great piles of logs, thousands on thousands, are ready for the spring flood. Before the ice has fairly ceased running, the logs are rolled into the water and the log drive is on its way. Some of the men are stationed along the shores to prevent the logs from lodging there, or floating into the bays or set-backs; some stand at the head of the bars or islands, where, with pike poles, they shove off the logs that might stop there and form a jam; others follow at the tail of the drive and clear up the shore of such pieces as may have drifted out of the current and been left behind. Then there is the cook, the most indispensable of all, who follows along the bank, pitching his tent from time to time in convenient places where the hungry crew can get their meals. When the freshet is subsiding, and the water falling so rapidly that the logs stick on every bar and along the shore, a splash dam is opened, and with the oncoming flood the work is resumed with all its interesting active scenes.

At times, in some crooked, rocky stream, a jam is formed, and thousands of logs are wedged fast in the channel, held back by some one log firmly braced against an impediment. Then occurs a thrilling scene, as the boss calls for volunteers to break the jam. There is always a prompt response. One or more daring fellows, impelled by pride in their work and love of applause, take their lives in their hands, as with axe and handspike they leap over the treacherous logs and place themselves at the head of the jam. Behind them are the thousands of logs, filling the angry stream from bank to bank, piled thickly to the bottom, some of them tossing, tumbling and leaping in air as the dammed up torrent forces them about in wild confusion. Beneath them is the swaying, rocking, unstable mass, in which is seen the log that forms the key to the position. The crew of drivers gather on the banks below the jam, where they watch with intense eagerness and anxiety the man who volunteers to cut or loosen this log. They note every stroke of the axe

F. A. VAN SANT, PHOTO.

ADIRONDACK LOG DRIVERS.

THE HALT AT NOON — COFFEE AND BAKED BEANS.

wielded by the hero as, coolly and undaunted, he proceeds with his work. The critical moment is close at hand. There is some more prying with the handspike, a few more blows with the axe, and the huge threatening mass begins to move. Above the sound of the foaming waters a loud shout of warning comes from the men below, and then, leaping from log to log as the jam breaks, the brave fellow reaches the shore in safety amid the applauding cheers of his comrades; or, as a cry of horror breaks from the watching crew, he loses his footing and disappears beneath the terrible grinding mass. Crushed and torn, a mere semblance of humanity, a man's body is found in the river later on, and in the camps next winter another chapter is added to the unwritten story of heroism as the men relate in quiet tones the story of the man who lost his life on the company's drive that spring. And so, amid scenes of toil and danger the work is done, the drive is safely inclosed in the big boom at the mills, and the job is completed.

There is probably no industry which involves so many varied details as the business of lumbering; none which requires so close attention at every stage of the work, and in which intelligence, strict economy, and, above all, thorough experience, are so necessary to profitable results. The failures in it have been so many that no one should undertake it who has not within him these specified requisites to success.

Modern Improvements.

It may be interesting to note here some of the more important changes that have taken place in the methods of lumbering since the Dutchman first set up his sawmill on Manhattan Island, most of which have occurred within the last century.

The chopper no longer uses a single bitted axe. In cutting down the tree a crosscut saw is used instead of an axe. The trunk of the fallen tree is sawed, not chopped, into logs of the required length. The forester in charge of a job insists that all trees shall be cut as close to the ground as possible, to save the timber which is lost in a high stump. On some of the jobs logs are skidded by wire ropes and steam power. Iced roads, easy grades, wide bunks and attention to details enable a team to haul a much larger load of logs. In some places water slides, miles in length, render log hauling unnecessary. Logs are placed on cars by steam loaders, lifted from the lake or stream by steam jack-works, and taken to the mill on a railroad instead of driving them down the river. The introduction of the planing mill* built up a new and distinct branch of the lumber business. In modern sawmills we find the oscillating gang, circular saws with inserted teeth,

* The Woodworth planing machine was patented in 1828, and the patent was extended twice.

mechanical appliances for handling logs, live rollers, and that most profitable of all improvements, the band saw.

Until recent years the word forest was seldom heard except as used rhetorically. It belonged to poetry and literature. The lumbermen and the people used the word woods instead. People lived in the backwoods, travelled through the woods, went into the woods, came out of the woods, or were lost in the woods; never, the forest. People spoke of the North woods, the South woods, the "nine-mile" woods, and the Shattygee (Chateaugay) woods.

But now we hear the words forest and forestry. And in chronicling the changes and improvements in the lumber industry of our State, mention must be made of the intelligent, conservative methods recently introduced in the management of our public and private woodlands by professional foresters whose working plans insure the preservation of our forests and perpetuation of our timber supply.

"Captains of Industry."

In recording the rise and progress of the lumber business it seems proper that some passing tribute, at least, should be paid to the memory of the men who, in one locality or another, were prominently identified with this important factor in the development of the wealth and resources of our State. They were men of both physical and mental vigor, possessed of sturdy virtues that made them respected not only in the communities where they lived, but wherever they were known. They had a keen sense of honor and fair dealing that made them known and described as men whose "word was as good as their bond," a common, homely expression, but one carrying praise that was well deserved. The grass has been growing on their graves for many years; but their memory is still cherished, their influence for good is still felt, and the world is better for their having lived in it.

Among the men, now dead and gone, who were prominently connected with the lumber business on the Upper Hudson, mention should be made of Abraham Wing,* James D. Weston, John J. Harris, Albert N. Cheney, Lewis L. Armes, Walter Geer, Orlin Mead, George Sanford, Orson Richards, Augustus Sherman, James Morgan, Charles H. Faxon, and Lemon Thomson.

---

* "Mr. Wing, born in 1791, had the foresight and judgment requisite for improving the golden chance by bringing to market the splendid pines with which the great Brant Lake Tract abounded. This rich and extensive lumber region, previously operated by the Fox Brothers, Alanson and Norman, had come into the possession of parties in Troy, who intrusted its management to Mr. Wing." (History of Queensbury. By A. W. Holden, M.D., Glens Falls, 1849.)

F. A. VAN SANT, PHOTO.

LOG SLIDE—J. & J. ROGERS CO.

FOR TRANSPORTING PULP LOGS TO THE AUSABLE RIVER, ESSEX CO., N. Y. THIS WATER-TROUGH IS SEVEN AND ONE-HALF MILES LONG.

On the Raquette River: Edward King, Ralph Pomeroy, Charles Pearson, George Richards, Matthias Vickery, James H. Carpenter, Edward W. Hutchings, Lyman H. Wilcox, Harrison Plummer, A. M. Adsit, and E. H. Rosekrans.

On the Tioga River: Deacon Simeon Hammond, Abijah Weston, William C. Bronson, William B. Stevens, and Julius Tremaine.

On the Saranac River: Senator Christopher F. Norton, Almon Thomas, James Hammond, David H. Parsons, Wales Parsons, Hartwell Brothers, C. A. Tefft, and Lorin Ellis.*

In Western New York: John D. Mersereau, of Portville, and the Weston Brothers, of Olean.

Woodpulp.

Within twenty years the logging industry in Northern New York has been materially affected by the demand for material necessary in the manufacture of woodpulp, an industry of comparatively recent development. Ground pulp, obtained by holding blocks of wood against a grindstone, was first made in this country in 1867, at Stockbridge, Mass. Chemical mills, in which the fibre is reduced by the action of acids under steam pressure, were introduced about the same time. Now, there are 293 mills, mechanical and chemical, in the United States, of which 102 are located in New York. Wisconsin comes next with 37; and Maine, with 30.

At first the New York mills used poplar only. This was deemed a desirable condition by our foresters, because this species does not appear to be available for any other purpose, while, at the same time, it is the one kind of tree with which Nature quickly reforests the burned areas in the Adirondacks. But poplar was soon discarded in favor of spruce, to which have been added within the last five years some of the other conifers, the process of manufacture having been so improved that a satisfactory fibre is now obtained from hemlock, pine, and balsam.

The effect on the timber cutting was soon evident. While the lumbermen formerly took nothing less than two-log trees, leaving nearly all that were twelve inches or less in diameter on the stump, the woodpulp men cut all the trees of certain species, large and small. This close cutting of the spruce and other kinds left no provision for future growth, and thinned the forest so severely in places

---

* For interesting and valuable facts relating to the history of lumbering operations in the Saranac Valley, see paper read by Hon. Everitt C. Baker before the Plattsburgh Institute, Jan. 14, 1901, and printed in the Plattsburgh Sentinel, Jan. 18, 1901.

that further damage was inflicted by winds and ice storms. The river drives which were composed formerly of fair sized logs were mixed thickly with slender butts; and on some streams where pulpwood only was running the drives were made up of four-foot lengths entirely.

The methods employed in a job for cutting pulp timber differ somewhat from those used in lumbering or getting out logs for sawmills. Very small trees as well as the large ones are cut, and when down are sawed into short lengths only four feet long, making the work of skidding, hauling, and river driving much easier. At first only the small trees were cut for pulpwood, the large timber being reserved for the sawmills and cut into logs of the usual length. But as the demand for woodpulp increased, the stumpage became more valuable for that purpose; all the spruce timber, both large and small, was cut. The largest spruce tree in the Adirondacks so far as known (41 inches in diameter on the stump) was cut for pulpwood, the shaft having being sawed into 22 short logs, each four feet long.

On some of the pulp jobs the timber is peeled in the woods before shipment, in order to save freight. The bark has no commercial value and is left in the woods where the peeling or "rossing" is done, the mass of dry bark strippings, which covers the ground thickly in places, increasing greatly the danger from fire. Much of the pulp timber in the Adirondacks is hauled direct to some railway station, and from there shipped to the mills, as under present market prices it will bear transportation a long distance. In other places the short pulp logs are driven down some stream and thence into a lake or pond near a railroad, where, by means of jack-works or conveyers, the sticks are lifted from the water and loaded on cars. In other localities a long haul by teams is avoided by the construction of water slides or wooden troughs, several miles in length, through which a shallow stream of water carries the sticks to the railroad, or to some river whence they are driven to the pulp mills, in the same manner as in a log drive.

In the vicinity of Benson Mines, St. Lawrence County, there is a water slide three miles long for conveying pulpwood to the railroad. This trough is 24 inches wide at the top and 10 at the bottom, with a depth of 20 inches. It is capable of moving 60 cords per hour. The company operating this slide had at one time a pile of pulpwood 1,000 ft. long, 26 ft. high, and 40 ft. wide, all of which had been transported from the woods to the railroad by this novel method. They had, also, an additional trough or slide in which sawed lumber was transported from the mill to the railroad by the same method. The J. & J. Rogers Pulp Company, of Ausable Forks, Essex County, N. Y., have on one of their jobs a water slide seven and one-half miles long, by which their pulp stock is carried to the Ausable River, and from there driven to their mills.

F. A. VAN SANT, PHOTO.

WATER SLIDE FOR TRANSPORTING LOGS.

ADIRONDACK FOREST, ESSEX CO., N. Y.

In 1898 the total cut of logs in the Adirondack forests amounted to 544,234,207 ft., of which 229,581,918 ft. were consumed in the pulp mills.

A peculiar effect of the woodpulp industry is the rise in value of spruce stumpage, which has increased beyond what the market value of the sawed lumber will warrant. Spruce stumpage is now worth so much more for woodpulp than for lumber that the sawmill men are unable to pay the price demanded for the standing timber; and unless there is some change in market conditions this species will not enter so largely hereafter into building operations, its place being taken to a great extent by hemlock or cheap pine.

Volume of Business.

The lumber industry of New York attained its maximum development at some time prior to 1865, when there were, according to the State census of that year, 3,963 sawmills. These included the little mills equipped with one saw only, none of which cut over 100,000 ft. in a year, and which composed perhaps three-fourths of the entire number.

From the tenth U. S. census, 1880, it appears that there were then 2,822 mills in New York, with an invested capital of $13,230,934, giving employment to 17,509 men, and paying out annually $2,162,972 in wages. The combined lumber product of these mills amounted in 1880 to 1,148,220,000 ft., board measure, not including lath, shingles, and staves.

Within the next twenty years there was a great decrease in the production, the timber supply having been exhausted with the exception of the Adirondack and Catskill forests. There are not 150 mills in the State to-day with an annual output of over 100,000 ft. The production is now confined almost wholly to the Adirondack region, the mills which are stocked from there having sawed in 1899 the following amounts:

	FEET B. M.
Spruce,	148,203,491
Hemlock,	46,545,772
Pine,	33,132,807
Hardwood,	24,296,554
	252,178,624
Shingles,	33,619,000
Lath,	49,329,090

To the amount of sawed lumber should be added 196,568,623 ft. of logs that went to the pulp mills, making the total forest output of Northern New York that year 447,747,247 ft.

There are several small sawmills in the Catskill counties, with a few others scattered through the western part of the State, their combined product not exceeding sixty million feet. The advocates of forest preservation and protection of our economic resources need no better argument than is contained in the figures showing the great decline in this industry within the last twenty years.

Lumber Markets of New York.

In addition to the ordinary lumber business connected with forest and sawmill there were in the State of New York great lumber markets or distributing points where the lumber was sold, not only the product of the State but immense shipments from Canada and the Northwest as well. The two principal markets were at Albany and Tonawanda.

Albany was the centre of a great lumber trade sixty years ago, and at one time surpassed all other points in the amount handled and volume of business. In 1872 there were forty-three wholesale firms, whose yards were grouped in the "lumber district," and who handled in the aggregate 660,000,000 ft. that year, their total sales amounting to over $15,000,000. Over 1,500 men were employed on the yards unloading and loading vessels, or in piling lumber, their total annual wages exceeding $600,000. But, owing to increased facilities for making direct, through shipments from the mills to the retailers, combined with unfavorable discriminations in freight rates, the business at Albany has declined so largely that the amount of lumber handled this year will not exceed 200,000,000 ft. The principal points of distribution for white pine are New York City, Long Island, Boston, New England, Newport, Fall River, Nantucket, Hudson River towns, West Indies, South America, Africa, Azores, and Australia. The shipments of spruce are confined mostly to Greater New York, Long Island and Hartford.

Tonawanda, unlike Albany, is a market in which all the lumber handled comes from outside the State — from the great pineries of the Northwestern States and Ontario. Still, some mention of it seems pertinent to this history on account of the immense lumber business carried on there. Next to Chicago and New York City it is the greatest lumber market in the United States or Canada. The entire stock received at this port is reshipped by rail or canal to other places, wherein it differs from Chicago and Greater New York, the latter places consuming a large proportion of their lumber receipts within their own limits.

F. A. VAN SANT, PHOTO.

TERMINUS OF WATER SLIDE.

LOGS FALLING INTO THE AUSABLE RIVER.

The business at Tonawanda commenced in 1857. The first cargo of lumber was shipped then from Canada by Brunson & Co. In 1865 it had become an important point in the general lumber business of the country, and its trade increased steadily until 1890, when it attained its maximum volume. In the latter year the receipts of lumber from the Great Lakes at this port amounted to 718,650,900 ft., to which may be added 13,039,600 lath and 52,232,300 shingles. Their combined values—lumber, lath and shingles—indicate a business that year of over $16,000,000. The number of persons employed—yardmen, planing mill hands, stevedores, and office men—is estimated at over 3,000, their annual wages exceeding in the aggregate $1,500,000. But the shipments have declined materially within the last ten years, the receipts in 1900 being reported at 396,429,483 ft. This decrease is due to through shipments from the West of carload lots direct to the retail yards in the Eastern States, and to disadvantages in freight rates. Still, the lumber business at Tonawanda is immense, employing a great many men, distributing a large amount of money annually in wages, and, as a distributing point for New England and the Middle States, contributing materially to the commercial supremacy of New York.

Oswego was also an important distributing point, where a large amount of Canadian lumber was handled and reshipped by canal. In 1870 the receipts of lumber at this port amounted to 289,315,329 ft.

The city of New York, owing to its export trade and large local demand, is also the centre of an immense lumber trade. In the year 1900 the total receipts of lumber from ocean, canal, river and railroad aggregated 1,246,014,604 ft.

The lumber industry in this State has been affected to some extent at different times by the tariff laws relating to Caandian imports. New York is not only a border State, but its closer proximity to most of the sawmills in Canada, combined with a larger number of entry ports and greater facilities for shipments, made its lumber markets more susceptible to foreign competition than those of any other State in the Union. Until recently the amount of lumber imported from Canada equaled or exceeded that produced in this State; and, owing to the number of lake ports, the Champlain route, and railroad transportation at various points on the border, the bulk of the importation found its way to the New York markets. In 1854, Hon. William L. Marcy, then Secretary of State at Washington, made a reciprocity treaty with Canada, under which lumber was admitted into the United States free of duty for ten years. Since 1865 the custom duties on sawed lumber have varied at times, and in some years were withdrawn altogether. The present tariff imposes a tax of $2.00 per M. on sawed lumber, which is equivalent, on an average, to an ad valorem duty of about thirteen per cent.* Under this impost the Canadian lumber imports have fallen off one-half within the last two years.

Such, in brief, is the history of lumbering operations and the lumber industry in the State of New York as brought down to the present time. It may be interesting to note here that, with the closing years of the century, improved methods of lumbering and a conservative system of forestry have been introduced in the woodlands of our State, which mark a distinct epoch in the history of this industry and promise to make the management of our forests a still more important and beneficent factor in the political economy of the State. Already some of the largest timber tracts belonging to private ownership in Northern New York are being lumbered under a system that not only insures immediate profit, but makes ample provision for forest preservation and a perpetual timber supply. The cutting is restricted not only to certain species, but to a fixed diameter that leaves a good proportion of the same species as the basis of another crop. Further provision for the future growth of merchantable timber is made by leaving at suitable intervals healthy specimens to serve as seed trees that will propagate a wind sown crop of seedlings in every opening. Economical methods of felling trees have been introduced which protect the young growth and, in addition, yield more timber per tree. New industries have arisen that are dependent on forest products, and utilize much of the material which heretofore has gone to waste.

The work is placed in charge of skilled foresters who mark each tree that is to be cut, and allow nothing cut that is not marked. The protective functions of the forest are carefully guarded, and no trees whatever are allowed to be cut on steep side hills or wherever a cutting might result in windfalls, soil erosion, or denudation.

The great primeval forests owned by the State have been carefully examined by competent foresters, who have made intelligent working plans under which the matured timber may be removed from time to time and a permanent, annual revenue secured to the State whenever the present constitutional restrictions are removed. And so, profiting by the lessons of the past, and encouraged by the successes of the present, the great lumber industry of the State enters upon another century of its existence with every promise that it will continne to add its full share to the honor and prosperity of the Commonwealth.

---

*The present tariff, enacted in 1897, puts a duty of $2.00 per M. feet on sawed pine, spruce, hemlock, balsam, maple, birch, beech, elm, ash, and walnut; and a duty of $1.00 per M. feet on whitewood, sycamore and basswood. Cedar, when sawed, is admitted at 15 per cent ad valorem. On planed lumber an additional charge is made of 50 cents per M. feet for each side dressed; and 50 cents more per M. for tongued and grooved boards of flooring.

F. A. VAN SANT, PHOTO.

CAUGHT NAPPING.

Appendix.

The Roll of Pioneer Lumbermen.

The Roll of Pioneer Lumbermen.

IT may fairly be assumed that in each locality throughout the State the commencement of the lumber industry was coincident with the erection of the first sawmill, all work before that being confined to individuals who labored to supply their personal needs only. But with the building of a mill there would ensue a combination of labor to supply a general demand, which constituted to some extent what is known as lumbering operations.

In the statistics appended here there will be found the year when the first settler located in each town mentioned, the list including the greater part of the State; also, the year in which the first sawmill was built in that town, together with the name of the pioneer who built it. It is understood that most of these towns had no existence then as such, but were set off or constituted subsequently; and the date here given means that the first settlement or erection of a sawmill occurred at that time in a locality which to-day is within the town named.

The historical facts thus given were compiled by the author from the various town and county records in the State Library at Albany. Much valuable information relating to the first settlement of each town was found in Hough's Gazeteer of the State of New York; and some interesting facts connected with colonial times were obtained from Dr. O'Callaghan's Documentary History of New York.

As this part of the work necessitated a careful examination of the early history of each one of the 977 towns in the State, it will be seen that the task required no small amount of time and patience. But the information thus obtained and set forth here will be of little interest to the general reader; and so the question may arise as to whether the result is worth the cost. In answer it should be said that any history of the lumber industry must properly commence with the beginning of that industry; and that there was no other way to determine definitely when it commenced in the various localities throughout the State except by ascertaining the date when the first sawmills were put in operation.

It may be noticed that the list is incomplete as regards some of the towns in the Hudson and Mohawk Valleys. An exhaustive examination was made of the town and colonial records relating to that region, but with only partial success, as but little mention is made in them of the erection of sawmills, or of the lumber industry.

These valleys, however, contained the oldest settlements, and from their borders the wave of civilization spread outward through the State, its advance being marked everywhere by the advent of the lumberman and his mill. Next, came the church and the schoolhouse.

The Beginnings of the Lumber Industry in the State of New York.

COUNTY.	TOWN.	FIRST SETTLEMENT.	FIRST SAWMILL.	BUILT BY.
Albany,	Albany,	1630	1654	Jacob Janse Flodder,
Albany,	Berne,	1750	1755	Jacob Weidman.
Albany,	Coeymans,	1673	1651*	Hans Jansen,
Albany,	Knox,	1789	1800	Amos Crary.
Albany,	New Scotland,	1700	1760	Uziah Conger.
Albany,	Watervliet,	1700	1775	Shaker Colony.
Albany,	Westerlo,	1763	1795	Lobdell & Baker,
Allegany,	Alfred,	1807	1821	E. S. Davis.
Allegany,	Allen,	1820	1820	Moses Treat.
Allegany,	Alma,	1833	1843	John W. Post.
Allegany,	Almond,	1796	1806	Phineas Stevens.
Allegany,	Amity,	1804	1806	Philip Church.
Allegany,	Andover,	1795	1819	Luther Strong.
Allegany,	Angelica,	1802	1803	Philip Church,
Allegany,	Belfast,	1803	1809	David Sanford.
Allegany,	Birdsall,	1816	1823	Hull & Peterson.
Allegany,	Bolivar,	1819	1822	Cowles Brothers.
Allegany,	Burns,†	1805	1813	David McCardy.
Allegany,	Caneadea,	1803	1816	John Hoyt.
Allegany,	Centerville,	1808	1813	Blanchard & Hotchkiss.
Allegany,	Clarksville,	1822	1832	Samuel King.
Allegany,	Cuba,	1814	1815	William Downer.

* The sawmill was built before there was any settlement in this town.

† The first deaths among the settlers in this town were those of Jeremiah Gregory, who was killed by the fall of a tree, April 4, 1812, and his twin brother, killed in the same manner on Sept. 17, same year.

BEGINNINGS OF THE LUMBER INDUSTRY — CONTINUED.

COUNTY.	TOWN.	FIRST SETTLEMENT.	FIRST SAWMILL.	BUILT BY.
Allegany,	Friendship,	1806	1815	Ebenezer Steenrod.
Allegany,	Genesee,	1823	1820	Newman Crabtree.
Allegany,	Granger,	1816	1819	Isaac Van Nostrand.
Allegany,	Grove,	1818	- -	John S. Culver.
Allegany,	Hume,	1807	1807	Roger Mills.
Allegany,	Independence,	1798	1800	John Cryder.
Allegany,	New Hudson,	1820	1829	James Davidson.
Allegany,	Rushford,	1808	1815	Matthew P. Cady.
Allegany,	Scio,	1805	1822	Benjamin Palmer.
Allegang,	Ward,	1817	1818	Stephen Easton.
Allegany,	Wellsville,	1801	1803	Nathaniel Dyke.
Allegany,	West Almond,	1816	1833	Enoch Hawks.
Allegany,	Willing,	1825	1829	Elijah Robinson.
Allegany,	Wirt,	1812	1824	Alvin Richardson.
Broome,	Barker,	1791	1801	Simeon Rogers.
Broome,	Chenango,	1784	1788	Henry French.
Broome,	Colesville,	1785	1792	Robert Harper.
Broome,	Conklin,	1788	1808	Robert Corbett.
Broome,	Fenton,	1788	1797	Elisha Pease.
Broome,	Lisle,	1791	1796	Edward Edwards.
Broome,	Sanford,	1787	1791	Nathan Dean.
Broome,	Vestal,	1785	1795	Bethias Du Bois.
Broome,	Windsor,	1786	1797	Nathan Lane.
Cattaraugus,	Allegany,	1820	1826	Reuben Lamberton.
Cattaraugus,	Carrollton,	1814	1826	Marcus Leonard.
Cattaraugus,	Cold Spring,	1820	1820	Crook & Basson.
Cattaraugus,	Conewango,	1816	1819	Sampson Crooker.*
Cattaraugus,	Dayton,	1810	1817	Silas Nash.
Cattaraugus,	East Otto,	1812	1823	Moses T. Beach.
Cattaraugus,	Ellicottville,	1815	1821	Orin Pitcher.

* Mrs. Crooker, wife of Sampson Crooker, killed a wildcat in her roost with a pair of tongs in the winter of 1819.

BEGINNINGS OF THE LUMBER INDUSTRY — CONTINUED.

COUNTY.	TOWN.	FIRST SETTLEMENT.	FIRST SAWMILL.	BUILT BY.
Cattaraugus,	Farmersville,	1817	1824	James Worden.
Cattaraugus,	Franklinville,	1806	1809	Henry Conrad.
Cattaraugus,	Freedom,	1811	1821	Enoch Howlett.
Cattaraugus,	Great Valley,	1812	1812	James Green.
Cattaraugus,	Hinsdale,	1806	1815	Lewis Wood.
Cattaraugus,	Humphrey,	1815	- -	Foster B. Salisbury.
Cattaraugus,	Ischua,	1812	1814	A. M. Farwell.
Cattaraugus,	Leon,	1819	1824	Ebenezer Collar.
Cattaraugus,	Little Valley,	1807	1809	David Powers.
Cattaraugus,	Lyndon,	1808	1843	Jason Sherman.
Cattaraugus,	Machias,	1813	1822	Andrew McBuzzell.
Cattaraugus,	Mansfield,	1817	1837	Clark Brothers.
Cattaraugus,	Napoli,	1818	1829	James Wait.
Cattaraugus,	New Albion,	1818	1836	Mathew Nealy.
Cattaraugus,	Olean,	1804	1807	Sheperd & Thrall.
Cattaraugus,	Otto,	1816	1822	Isaac W. Sherman.
Cattaraugus,	Perrysburgh,	1815	- -	Isaac Balcom.
Cattaraugus,	Persia,	1811	1814	Ahaz Allen.
Cattaraugus,	Portville,	1805	1807	Green & Dodge.
Cattaraugus,	Randolph,	1820	1823	Thomas Harvey.
Cattaraugus,	South Valley,*	1798	1801	Quaker Colony.
Cattaraugus,	Yorkshire,	1810	1814	Isaac Williams.
Cayuga,	Brutus,	1800	1808	Lewis Putnam.
Cayaga,	Conquest,†	1800	1808	——— Twitchell.
Cayuga,	Sterling,	1805	1817	John Cooper.
Cayuga,	Throop,	1790	1798	Prentice Palmer.
Chautauqua,	Arkwright,	1807	1818	Benjamin Orton.
Chautauqua,	Carroll,	1807	1811	John Frew.

*The mills did work for the white settlers on the usual terms, and furnished lumber for the Indians free.

†In 1804 James Perkins built the first framed house, sawing out the whole lumber with a whipsaw. The building was still standing in 1859, a monument of persevering industry.

J. Y. MC CLINTOCK, PHOTO.

MODERN BAND SAW.

NORWOOD MANUFACTURING COMPANY, TUPPER LAKE, N. Y.

BEGINNINGS OF THE LUMBER INDUSTRY — CONTINUED.

COUNTY.	TOWN.	FIRST SETTLEMENT.	FIRST SAWMILL.	BUILT BY.
Chautauqua,	Charlotte,	1809	1810	Samuel Sinclair.
Chautauqua,	Cherry Creek,	1812	1824	William Kilbourn.
Chautauqua,	Ellery,	1806	1808	William Bemus.
Chautauqua,	Ellicott,	1806	1808	Edward Works.
Chautauqua,	Gerry,	1811	1819	Hines & Newton.
Chautauqua,	Hanover,	1797	1804	Abel Cleveland.
Chautauqua,	Harmony,	1805	1810	Reuben Slayton.
Chautauqua,	Kiantone,	1807	- -	Robert Russell.
Chautauqua,	Mina,	1816	1824	Alex. Finley.
Chautauqua,	Poland,	1805	1805	Dr. Thos. R. Kennedy.
Chautauqua,	Pomfret,	1804	1807	Baker, Berry & Co.
Chautauqua,	Villenova,	1810	1815	John Kent.
Chautauqua,	Westfield,	1801	1804	John McMahan.
Chemung,	Ashland,	1788	1800	Isaac Baldwin.
Chemung,	Baldwin,	1813	1828	Elisha Hammond.
Chemung,	Big Flats,	1787	1795	William Miller.
Chemung,	Catlin,	1816	1827	James Wheeler.
Chemung,	Chemung,	1786	1790	Major Wm. Wynkoop.*
Chemung,	Erin,	1815	1824	McMillan Brothers.
Chemung,	Horseheads,	1787	1805	Nathan Teal.
Chemung,	Southport,	1788	1798	Col. Abraham Miller.
Chemung,	Van Etten,	1795	1800	Isaac Swartwood.
Chemung,	Veteran,	1775	1805	——— Teal.
Chenango,	Afton,	1786	- -	Cooper & Miner.
Chenango,	Columbus,	1791	1794	Job Vail.
Chenango,	Greene,	1792	1795	Conrad Sharp.
Chenango,	Lincklaen,	1796	- -	Catlin & Shipman.
Chenango,	Macdonough,	1795	1798	Henry Ludlow.
Chenango,	New Berlin,	1790	- -	Job Vail.
Chenango,	Norwich,	1788	-	Elisha Smith.

* Major Wynkoop built the first frame house in this town, the boards and timbers for which were sawed out with a whipsaw.

BEGINNINGS OF THE LUMBER INDUSTRY—CONTINUED.

COUNTY.	TOWN.	FIRST SETTLEMENT.	FIRST SAWMILL.	BUILT BY
Chenango,	Otselic,	1800	- -	James Rush.
Chenango,	Smithville,	1797	1805	Timothy Scoville.
Clinton,	Altona,	1800	1819	Benjamin Mooers.
Clinton,	Ausable,	1794	1806	Thaddeus Mason.
Clinton,	Black Brook,	1823	1825	Burt & Vanderwarker.
Clinton,	Champlain,	1789	1787	Lieut. Pliny Moore.
Clinton,	Chazy,	1783	1801	Atwood Brothers.
Clinton,	Clinton,	1818	1850	John McCoy.
Clinton,	Ellenburgh,	1800	1824	John R. Murray.
Clinton,	Mooers,	1796	1804	George Perry.
Clinton,	Peru,	1785	1810	John Cochran.
Clinton,	Plattsburgh,	1765	1769	Count de Freydenburgh.
Clinton,	Saranac,	1802	1806	Isaiah Ferris.
Clinton,	Schuyler Falls,	1794	1801	Ezra Turner.
Columbia,	Canaan,	1766	1774	Col. Wm. B. Whiting.
Columbia,	Kinderhook,	1661	1665	Frank Pieters Clavers.
Columbia,	Livingston,	1708	1710	Robert Livingston.
Cortland,	Marathon,	1794	- -	John Hunt.
Cortland,	Taylor,	1794	1816	Rockwell Brothers.
Cortland,	Virgil,	1792	1801	Daniel Edward.
Cortland,	Willett,	1793	- -	Benjamin Wilson.
Delaware,	Andes,	1770	1811	John Vaughan.
Delaware,	Bovina,	1792	1802	Matthew Russell.
Delaware,	Colchester,	1774	1790	William Rose.
Delaware,	Davenport,	1786	1792	Daniel Prentice.
Delaware,	Delhi,	1784	1800	Oliver Peake.
Delaware,	Franklin,	1784	1806	Samuel Hutchinson.
Delaware,	Hamden,	1779	1799	Peake & Ward.
Delaware,	Harpersfield,	1776	- -	Col. John Harper.
Delaware,	Kortright,	1773	1780	Ezekiel Johnson.
Delaware,	Masonville,	1792	1799	Joseph Bushnell.
Delaware,	Middletown,	1763	1790	Benjamin Akerly.

BEGINNINGS OF THE LUMBER INDUSTRY—CONTINUED.

COUNTY.	TOWN.	FIRST SETTLEMENT.	FIRST SAWMILL.	BUILT BY.
Delaware,	Roxbury,	1789	1805	——— Ferris.
Delaware,	Sidney,	1772	1790	——— Carr.
Delaware,	Stamford,	1790	1795	Joseph Warn.
Delaware,	Tompkins,	1780	1786	Jesse Dickinson.*
Delaware,	Walton,	1785	- -	M. Goodrich.
Dutchess,	Rhinebeck,	1700	1740	Jacob Rutsen.
Erie,	Alden,	1810	1814	John C. Rogers.
Erie,	Amherst,	1800	1801	John Thompson.
Erie,	Aurora,	1804	1806	Phineas Stephens.
Erie,	Brant,	1818	1822	Samuel Butts.
Erie,	Cheektowaga,	1808	1810	Samuel Le Suer.
Erie,	Clarence,	1799	1804	Asa Ransom.
Erie,	Colden,	1810	1811	Richard Buffum.
Erie,	Collins,	1808	1809	Quaker Colony.
Erie,	Concord,	1807	1812	Rufus Eaton.
Erie,	East Hamburgh,	1804	1803	David Eddy.
Erie,	Eden,	1808	1811	Elisha Welch.
Erie,	Elma,	1829	1832	——— Estabrook.
Erie,	Evans,	1804	1815	——— ———.
Erie,	Holland,	1807	1815	Ephraim Woodruff.
Erie,	Marilla,	1829	1828	Jesse Barton.
Erie,	Sardinia,	1809	1812	Sumner Warren.
Essex,	Chesterfield,	1792	1802	Robert Hoyle.
Essex,	Crown Point,	1804	1810	Allen Penfield.
Essex,	Elizabethtown,	1792	1814	Amos Rice.
Essex,	Essex,	1783	1784	Daniel Ross.
Essex,	Jay,†	1796	1798	William Mallory.

* This mill, soon after its erection, was carried away by the famous "punkin flood" that inundated the valleys of the Susquehanna and its tributaries in the fall of 1787. The corn fields were swept bare, and the yellow pumpkins that thickly dotted the surface of the swollen streams were so conspicuous that the descriptive name just mentioned still survives among the household words in Southern New York.

† "In the vicinity of Upper Jay the lumber business was killed as early as 1820 by the girdling of all the trees to facilitate the clearing of the land." (Hist. Essex Co. by H. P. Smith, 1885. Syracuse: D. Mason & Co.)

BEGINNINGS OF THE LUMBER INDUSTRY — CONTINUED.

COUNTY.	TOWN.	FIRST SETTLEMENT.	FIRST SAWMIL.	BUILT BY.
Essex,	Lewis,	1798	1809	Asa Farnsworth.
Essex,	Minerva,	1804	1807	William Hill.
Essex,	Moriah,	1785	1810	Alexander Spencer.
Essex,	North Hudson,	1802	1812	Elihu Phelps.
Essex,	Schroon,	1797	1814	Joseph Richards.
Essex,	Ticonderoga,*	1769	1772	Fox & Huntington.
Essex,	Willsborough,	1765	1767	William Gilliland.
Essex,	Wilmington,	1803	1812	Reuben Sanford.
Franklin,	Bellmont,	1816	1825	Roswell A. Weed.
Franklin,	Chateaugay,	1796	1797	David Mallory.
Franklin,	Constable,	1800	1803	James Welch.
Franklin,	Dickinson,	1810	1830	Warren Ives.
Franklin,	Duane,	1823	1823	James Duane.
Franklin,	Fort Covington,	1794	1796	William Gray.
Franklin,	Franklin,	1827	1827	McLenatham & Wells.
Franklin,	Malone,	1802	1804	N. & J. Wood.
Franklin,	Moira,	1803	1803	Appleton Foote.
Franklin,	Westville,	1800	1801	Amos Welch.
Fulton,	Broadalbin,	1783	1810	Duncan McMartin.
Fulton,	Caroga,	1785	1790	Cornelius Van Allen.
Fulton,	Ephratah,	1743	1808	Henry Yanney.
Fulton,	Johnstown,	1760	1762	Sir Wm. Johnson.
Fulton,	Mayfield,	1761	1773	Sir Wm. Johnson.
Fulton,	Oppenheim,	1791	1806	Henry Cline.
Fulton,	Stratford,	1799	1806	Martin Nichols.
Genesee,	Alabama,	1806	1824	Samuel Whitcomb.
Genesee,	Alexander,	1802	1804	Rea & Fellows.
Genesee,	Batavia,	1801	1801	Joseph Ellicott.
Genesee,	Bergen,	1805	1811	Jared Merrill.
Genesee,	Byron,	1807	1813	William Shepherd.

* The French troops, while engaged in the construction of Fort Carillon, built a sawmill at the outlet of Trout Brook; but it was destroyed soon after.

J. Y. MC CLINTOCK, PHOTO.

CIRCULAR SAW.

NORWOOD MFG. CO., TUPPER LAKE, FRANKLIN CO., N. Y.

BEGINNINGS OF THE LUMBER INDUSTRY—CONTINUED.

COUNTY.	TOWN.	FIRST SETTLEMENT.	FIRST SAWMILL.	BUILT BY.
Genesee,	Darien,	1803	1809	Amos Humphrey.
Genesee,	Elba,	1801	1810	Horace Gibbs.
Genesee,	Oakfield,	1801	1811	Christopher Kenyon.
Genesee,	Pembroke,	1804	1808	Samuel Carr.
Genesee,	Stafford,	1801	1810	Amos Stow.
Greene,	Ashland,	1785	1820	Marshall Lewis.
Greene,	Cairo,	1760	1808	Enoch Hyde.
Greene,	Catskill,*	1650	1684	Dirk Tenisse Van Vechten.
Greene,	Coxsackie,	1695	1750	Casparus Bronk.
Greene,	Durham,	1770	1788	Jared Smith.
Greene,	Greenville,	1750	1800	Nathaniel Holmes.
Greene,	Halcott,	1800	1820	Henry Hosford.
Greene,	Hunter,	1800	1820	Roger Bronson.
Greene,	Jewett,	1783	1795	Laban Andrews.
Greene,	Lexington,	1777	1824	John Bray.
Greene,	New Baltimore,	1785	1800	Charles Titus.
Greene,	Prattsville,	1712	1823	——— Smedburg.
Greene,	Windham,	1817	1817	Jared Clark.
Hamilton,	Arietta,	1827	1831	——— ———.
Hamilton,	Benson,	1825	1826	Nathaniel Lobdell.
Hamilton,	Hope,	1790	1803	——— ———.
Hamilton,	Indian Lake,	1843	1849	Wing Lumber Company.
Hamilton,	Lake Pleasant,	1795	1795	——— Foster.
Hamilton,	Long Lake,	1830	1836	E. H. St. John.†
Hamilton,	Morehouse,	1833	1833	Andrew K. Morehouse.
Hamilton,	Wells,	1798	1829	Halsey Rogers.
Herkimer,	Danube,	1730	1799	Samuel Haupt.
Herkimer,	Fairfield,	1770	1781	Samuel Green.

*A grist and sawmill combined. Robert Livingston, in a letter written by him in 1712, says, "A little mill at Catskill grinds so coarse it cannot be bolted."

†He built it for a landowner named Hammond, receiving in payment a stated sum of money and five lots of land (800 acres?). Under the terms of the contract he also cut out the first road from Newcomb to Long Lake.

BEGINNINGS OF THE LUMBER INDUSTRY — CONTINUED.

COUNTY.	TOWN.	FIRST SETTLEMENT.	FIRST SAWMILL.	BUILT BY.
Herkimer,	Frankfort,	1775	1794	John Hollister.
Herkimer,	Litchfield,	1789	1806	——— Talcott.
Herkimer,	Newport,	1791	1793	Benjamin Bowen.
Herkimer,	Norway,	1786	1793	Capt. David Hinman.
Herkimer,	Russia,	1792	1797	Benjamin Hinman,
Herkimer,	Stark,	1775	1776	Abraham Van Horne.
Herkimer,	Webb,*	1799	1800	Gov. John Brown.
Herkimer,	Wilmurt,	1790	1790	Arthur Noble.
Herkimer,	Winfield,	1792	1795	Joseph Walker.
Jefferson,	Adams,	1800	1802	David Smith.
Jefferson,	Antwerp,	1803	1806	Silas Ward.
Jefferson,	Brownville,	1799	1800	Gen. Jacob Brown.
Jefferson,	Champion,	1797	1799	Joel Mix.
Jefferson,	Clayton,	1802	1804	Smith & Delamater.
Jefferson,	Ellisburg,	1797	1797	Lyman Ellis.
Jefferson,	Henderson,	1799	1807	Willis Fellows.
Jefferson,	Hounsfield,	1800	1802	Augustus Sacket.
Jefferson,	Le Ray,	1801	1802	Benjamin Brown.
Jefferson,	Lorraine,	1802	1804	——— Frost.
Jefferson,	Lyme,	1801	1803	——— ———.
Jefferson,	Orleans,	1806	1816	Dr. Andrus.
Jefferson,	Philadelphia,	1804	1805	Thomas and John Townsend.
Jefferson,	Rodman,	1801	1804	William Rice.
Jefferson,	Rutland,	1799	1802	David Coffeen.
Jefferson,	Theresa,	1810	1810	James D. Le Ray.
Lewis,	Croghan,	1828	1830	Somerville Stewart.
Lewis,	Denmark,	1800	1801	Nathan Munger.
Lewis,	Diana,	1830	1833	Faskit Harris.
Lewis,	High Market,	1814	- -	James McVicker.
Lewis,	Lewis,	1799	1800	Joel Jenks.

* This mill was built at Old Forge by Gov. John Brown, of Providence, R. I., the owner of Brown's Tract in the Adirondacks.

BEGINNINGS OF THE LUMBER INDUSTRY — CONTINUED.

COUNTY.	TOWN.	FIRST SETTLEMENT.	FIRST SAWMILL.	BUILT BY.
Lewis,	Leyden,	1794	1795	Bela Butterfield.
Lewis,	Lowville,	1797	1798	Daniel Kelley.
Lewis,	Martinsburgh,	1801	1803	Walter Martin.
Lewis,	Montague,	1846	1848	S. P. Sears.
Lewis,	New Bremen,	1821	1826	Charles Dayan.
Lewis,	Osceola,	1838	1841	William Roberts.
Lewis,	Turin,	1797	1799	Eleazer House.
Lewis,	Watson,	1800	1801	Isaac Puffer.
Lewis,	West Turin,	1795	1796	Nathaniel Shaler.
Livingston,	Avon,	1785	1790	Timothy Hosmer.
Livingston,	Conesus,	1794	1804	——— ———.
Livingston,	Leicester,	1789	1792	Ebenezer Allen.
Livingston,	Lima,	1788	1796	Reuben Thayer.
Livingston,	Livonia,	1792	1795	——— Higby.
Livingston,	North Dansville,	1795	1796	David Scholl.
Livingston,	Nunda,	1806	1818	Willoughby Lovell.
Livingston,	Ossian,	1804	1809	Nathaniel Porter.
Livingston,	Portage,	1810	1816	Russell Messenger.
Livingston,	Springwater,	1807	1809	Samuel Hines.
Madison,	Cazenovia,	1793	1794	John Lincklaen.
Madison,	De Ruyter,	1793	1807	Joseph Rich.
Madison,	Eaton,	1793	1795	Joshua Leland.
Madison,	Georgetown,	1804	1806	Mitchell Atwood.
Madison,	Lebanon,	1792	- -	Elisha Wheeler.
Madison,	Madison,	1793	1793	Henry W. Bond.
Madison,	Nelson,	1794	1800	Jeremiah Clark.
Madison,	Smithville,	1795	1801	Peter Smith.
Madison,	Stockbridge,	1791	1794	Stockbridge Indians.
Madison,	Sullivan,	1790	- -	John G. Moyer.
Monroe,	Greece,	1792	1810	Nathaniel Jones.
Monroe,	Henrietta,	1806	- -	Jonathan Smith.
Monroe,	Ogden,	1802	1811	William H. Spencer.

BEGINNINGS OF THE LUMBER INDUSTRY — CONTINUED.

COUNTY.	TOWN	FIRST SETTLEMENT.	FIRST SAWMILL.	BUILT BY.
Monroe,	Parma,	1794	1811	Jonathan Whitney.
Monroe,	Penfield,	1801	- -	Daniel Penfield.
Monroe,	Riga,	1805	1808	Samuel Church.
Monroe,	Webster,	1805	1806	Caleb Lyon.
Monroe,	Wheatland,	1789	1810	Peter Shaeffer, Jr.
Montgomery,	Amsterdam,	1716	1742	Sir William Johnson.
Montgomery.	Canajoharie,	1770	1770	Col. Hendrick Frey.
Montgomery,	Charleston,	1737	1785	Judah Burton.
Montgomery,	Florida,	1710	1750	Philip Frederick.
Montgomery,	Glen,	1705	1790	Peter Quackenboss.
Montgomery,	Minden,	1750	1740	——— Fox.
Montgomery,	Root,	1770	- -	Solomon Hamilton.
New York,	Manhattan,	1614	1633	West India Company.
Niagara,	Cambria,	1800	1806	Joseph Hewett.
Niagara,	Lewiston,	1800	1808	Joseph Howell.
Niagara,	Newfane,	1807	1811	James Van Horn.
Niagara,	Porter,	1803	1816	John Clapsaddle.
Niagara,	Royalton,	1803	1817	Gad Warner.
Niagara,	Somerset,	1810	1822	John Randolph.
Niagara,	Wheatfield,	1802	1825	Col. John Sweeney.
Niagara,	Wilson,	1810	1815	Daniel Sheldon.
Oneida,	Augusta,	1793	1795	T. Cassaty.
Oneida,	Ava,	1798	1801	Benjamin Jones.
Oneida,	Boonville,	1795	1796	Holland Land Co.
Oneida,	Bridgewater,	1788	- -	Major Farwell.
Oneida,	Camden,	1796	1798	Jesse Curtis.
Oneida,	Lee,	1790	1796	David Smith.
Oneida,	Manchester,	1787	1788	Captain Casey.
Oneida,	Marcy,	1793	1825	John F. Allen.
Oneida,	Vienna,*	1798	1801	Ambrose Jones.

* The first death in the town occurred by an accident in a sawmill, in 1801, whereby Alex. Graves was killed.

J. Y. MC CLINTOCK, PHOTO.

JACK-LADDER AND ENDLESS CHAIN.

FOR HAULING LOGS OUT OF THE POND INTO THE MILL.

BEGINNINGS OF THE LUMBER INDUSTRY — CONTINUED.

COUNTY.	TOWN.	FIRST SETTLEMENT.	FIRST SAWMILL.	BUILT BY.
Oneida,	Westmoreland,	1786	1790	Jonathan Dean.*
Onondaga,	Camillus,	1790	1806	Munro & Wheeler.
Onondaga,	Cicero,	1790	1823	Freeman Hotchkiss.
Onondaga,	Clay,	1793	1811	Abraham Young.
Onondaga,	Dewitt,	1789	1792	Asa Danforth.
Onondaga,	Elbridge,	1793	1797	William Stevens.
Onondaga,	Fabius,	1794	1800	Thomas Miles.
Onondaga,	Geddes,	1794	1825	Noah Smith.
Onondaga,	Lafayette,	1791	1795	James Sherman.
Onondaga,	Lysander,	1793	1807	Dr. Jonas C. Baldwin.
Onondaga,	Manlius,	1790	1793	Elijah Phillips.
Onondaga,	Marcellus,	1794	1796	Bradley & Rice.
Onondaga,	Onondaga,	1787	1793	Turner Fenner.
Onondaga,	Pompey Hill,	1792	1796	Pratt & Smith.
Onondaga,	Skaneateles,	1793	1796	Jedediah Sanger.
Onondaga,	Spafford,	1794	1810	Josiah Walker.
Onondaga,	Tully,	1795	1810	Peter Van Camp.
Onondaga,	Van Buren,	1792	- -	Skeels & Paddock.
Ontario,	Canadice,	1807	- -	John Algur.
Ontario,	East Bloomfield,	1789	1790	Gen. Fellows.
Ontario,	Farmington,	1789	1795	Smith Brothers.
Ontario,	Gorham,	1789	1807	Buckley & Craft.
Ontario,	Naples,	1790	1792	Clark & Metcalf.
Ontario,	Phelps,	1789	1795	Seth Dean.
Ontario,	Richmond,	1789	1795	Thomas Morris.
Ontario,	Seneca,	1787	1793	P. B. Wisner.
Ontario,	South Bristol,	1789	1795	Gamaliel Wilder.
Ontario,	Victor,	1789	1792	E. & J. Boughton.
Ontario,	West Bloomfield,	1789	1798	Ebenezer Curtis.
Orange,	Chester,	1751	1810	Richard Bull.

*A MSS. account of Indian mythology written by Mr. Dean is in the State Library, Albany, N. Y.

BEGINNINGS OF THE LUMBER INDUSTRY — CONTINUED.

COUNTY.	TOWN.	FIRST SETTLEMENT.	FIRST SAWMILL.	BUILT BY.
Orange,	Crawford,	1740	1751	Johannes Snyder.
Orange,	Montgomery,	1722	1768	Robert Milliken.
Orange,	Newburgh,	1709	1784	Capt. Thos. Machin.
Orange,	New Windsor,	1685	1728	Samuel Hazard.
Orange,	Wallkill,	1767	1776	——— Carpenter.
Orange,	Warwick,	1719	1760	Daniel Burt.
Orange,	Wawayanda,	1738	1760	Isaac Finch.
Orleans,	Barre,	1811	1816	William White.
Orleans,	Clarendon,	1811	1811	Eldridge Farwell.
Orleans,	Gaines,	1808	1813	Henry Drake.
Orleans,	Kendall,	1812	1819	Auger & Boyden.
Orleans,	Ridgeway,	1809	1805	Holland Land Company.
Orleans,	Shelby,	1810	1812	Joseph Ellicott.
Oswego,	Albion,	1812	1813	Lilly Brothers.
Oswego,	Amboy,	1805	1822	Joseph Murphy.
Oswego,	Boylston,	1810	1822	Reuben Snyder.
Oswego,	Constantia,	1790	1795	George Scriba.
Oswego,	Granby,	1792	1814	Schenck & Wilson.
Oswego,	Hannibal,	1802	1811	Silas Crandell.
Oswego,	Mexico,	1798	- -	George Scriba.
Oswego,	New Haven,	1798	1805	Ira Foot.
Oswego,	Orwell,	1806	1810	Joseph Watson.
Oswego,	Palermo,	1806	1812	Phineas Chapin.
Oswego,	Parish,	1804	1807	Way & Allen.
Oswego,	Redfield,	1798	1801	——— ———.
Oswego,	Richland,	1801	1806	John Hoar.
Oswego,	Sandy Creek,	1804	1805	William Skinner.
Oswego,	Schroeppel,	1800	1819	H. W. Schroeppel.
Oswego,	Volney,	1793	1796	——— Goodell.
Oswego,	Williamstown,	1801	1803	Isaac Alden.
Otsego,	Maryland,	1793	1795	Jotham Boughton.
Otsego,	Middlefield,	1775	1776	Alexander McCollum.

BEGINNINGS OF THE LUMBER INDUSTRY — CONTINUED.

COUNTY.	TOWN.	FIRST SETTLEMENT.	FIRST SAWMILL.	BUILT BY.
Otsego,	Milford,	1770	1792	Matthew Cully.
Otsego,	Roseboom,	1800	1806	Abram Roseboom.
Otsego,	Springfield,	1762	1775	Garret Staats.
Otsego,	Westford,	1790	1796	Artemas Howe.
Otsego,	Worcester,	1789	1791	Silas Crippen.
Putnam,	Carmel,	1739	1750	——— Kellogg.
Putnam,	Kent,	1754	1783	Elisha Cole.
Putnam,	Philipstown,	1730	1762	Col. Beverly Robinson.
Putnam,	Putnam Valley,	1740	1785	Isaac Post.
Putnam,	Southeast,	1730	1740	Col. Jonathan Crane.
Queens,	Jamaica,	1656	1675	Joseph Carpenter.*
Queens,	Oyster Bay,	1653	1673	Henry Townsend.†
Rensselaer,	Berlin,	1765	1780	Amos Sweet.
Rensselaer,	Grafton,	1786	1799	Josiah Litchfield.
Rensselaer,	Sandlake,	1766	1791	Solomon Taylor.
Rensselaer,	Stephentown,	1765	1800	——— Younglove.
Rensselaer,	Troy City,	1659	1663	Jan Barentson Wemp.
Richmond,	Castleton,	1640	1669	John Palmer.
Rockland,	Ramapo,	1740	1795	John Suffern.
St. Lawrence,	Brasher,	1814	1815	G. B. R. Gove.
St. Lawrence,	Canton,	1800	1802	Stillman Foote.
St. Lawrence,	Colton,	1824	1825	Horace Garfield.
St. Lawrence,	DeKalb,	1803	1809	Charles Boreland.
St. Lawrence,	Edwards,	1812	1824	Job Winslow.
St. Lawrence,	Fine,	1823	1828	James C. Haile.

*Joseph Carpenter and Caleb Carman entered into an agreement with the town, whereby they were permitted to use timber from the common lands, "except clapboard and rayle trees under eighteen inches," and were to saw for the town "twelve pens in the hundred cheaper than any other person of any other town;" and for citizens of the town "that bringeth the timber, one halfe of the sawn stuff for their laboure, provided that it is only for their owne use."

†For building this mill the town granted to Townsend, and his heirs forever, the right to cut and use timber from any part of the town he should choose; also, the right to sell such timber, either in the town or out of it.

BEGINNINGS OF THE LUMBER INDUSTRY — CONTINUED.

COUNTY.	TOWN.	FIRST SETTLEMENT.	FIRST SAWMILL.	BUILT BY.
St. Lawrence,	Fowler,	1807	1808	James Haile.
St. Lawrence,	Gouverneur,	1806	1809	Lewis R. Morris.
St. Lawrence,	Hermon,	1812	1818	Milton Johnson.
St. Lawrence,	Hopkinton,	1803	1824	Isaac R. Hopkins.
St. Lawrence,	Lawrence,	1807	1809	Ephraim Martin.
St. Lawrence,	Lisbon,	1800	1804	D. W. Church.
St. Lawrence,	Louisville,	1800	1805	Asa Day.
St. Lawrence,	Macomb,	1817	1818	Timothy Pope.
St. Lawrence,	Madrid,	1801	1803	Seth Roberts.
St. Lawrence,	Massena,	1792	1792	Amable Foucher.
St. Lawrence,	Norfolk,	1809	1810	Timothy W. Osborn.
St. Lawrence,	Ogdensburg,*	1749	1751	Father Picquet.
St. Lawrence,	Oswegatchie,	1796	1797	Nathan Ford.
St. Lawrence,	Parishville,	1810	1810	Barnes Brothers.
St. Lawrence,	Pierrepont,	1807	1819	Cox & Dimmick.
St. Lawrence,	Pitcairn,	1824	1828	P. Jenny.
St. Lawrence,	Potsdam,	1803	1803	Benjamin Raymond.
St. Lawrence,	Rossie,	1807	1810	D. W. Church.
St. Lawrence,	Russell,	1805	1805	Joel Clark.
St. Lawrence,	Stockholm,	1802	1804	Samuel Reynolds.
Saratoga,	Charlton,	1774	1783	John Roger.
Saratoga,	Greenfield,	1784	1789	Gershom Morehouse.
Saratoga,	Hadley,	1783	1791	Delane & Hazard.
Saratoga,	Half Moon,	1690	1762	——— ———.
Saratoga,	Northumberland	1775	1777	——— Munroe.
Saratoga,	Providence,	1775	1786	——— Corey.

*In an official report made to the Canadian Parliament in 1851, entitled "Titles and documents relating to the seignorial tenure," there is a copy of the grant made to Abbe Picquet giving permission to build a sawmill. This concession, signed by Francois Bigot, the intendant at Quebec, states "that for the usefulness of the said mill, it is necessary that there should be attached thereto a tract of land on which to receive the saw logs as well as the boards and other lumber," and grants for this purpose "land of one arpent and a half in front, by the same depth." This land now forms part of the city of Ogdensburg.

BEGINNINGS OF THE LUMBER INDUSTRY—CONTINUED.

COUNTY.	TOWN.	FIRST SETTLEMENT.	FIRST SAWMILL.	BUILT BY.
Saratoga,	Wilton,	1774	1784	John Laing.
Schenectady,*	———,	- -	- -	——— ———.
Schoharie,	Broome,	1791	1794	Griswold & Wells.
Schoharie,	Cobleskill,	1750	1774	Christian Brown.
Schoharie,	Jefferson,	1794	1796	Stephen Judd.
Schoharie,	Richmondville,	1770	- -	Company of settlers.
Schoharie,	Seward,	1754	1773	William Hynds.
Schoharie,	Sharon,	1771	1784	John Hutt.
Schoharie,	Summit,	1794	1798	——— Van Buren.
Schoharie,	Wright,	1771	1783	Zimmer & Becker.
Schuyler,	Catharine,	1788	1791	Phineas Bowers.
Schuyler,	Cayuta,	1798	1816	Jesse D. White.
Schuyler,	Dix,	1798	1828	Col. Green Bennett.
Schuyler,	Hector,	1790	1795	Reuben Smith.
Schuyler,	Orange,	1802	- -	Wm. Conlogue.
Schuyler,	Reading,	1790	- -	Eliadia Parker.
Schuyler,	Tyrone,	1799	1801	——— ———.
Seneca,	Fayette,	1789	1797	Samuel Bear.
Seneca,	Seneca Falls,	1787	1795	Wilhelmus Mynderse.
Seneca,	Tyre,	1794	1807	Nicholas Traver.
Steuben,	Addison,	1791	1793	George Goodhue.
Steuben,	Avoca,	1800	1809	Henry Kennedy.
Steuben,	Bath,	1793	1793	Chas. Williamson.
Steuben,	Bradford,	1793	1795	Frederick Bartles.
Steuben,	Cameron,	1800	1808	Richard Hadley.
Steuben,	Campbell,	1800	1807	Campbell & Stephens.
Steuben,	Caton,	1819	1822	Abner Gilbert.
Steuben,	Cohocton,	1796	1808	Jonas Cleland.
Steuben,	Corning,	1788	1793	Ichabod Patterson.

* The colonial patent of 1684, embracing lands in this county and the present site of the city of Schenectady, refers to sawmills which had already been built within the territory granted; but nothing appears now in the town or county records to show when or by whom they were erected.

BEGINNINGS OF THE LUMBER INDUSTRY — CONTINUED.

COUNTY.	TOWN.	FIRST SETTLEMENT.	FIRST SAWMILL.	BUILT BY.
Steuben,	Dansville,	1804	1816	Robert Fuller.
Steuben,	Erwin,	1787	1820	Samuel Erwin.
Steuben,	Fremont,	1812	1816	Daniel Upson.
Steuben,	Hartsville,	1809	1827	Daniel P. Carpenter.
Steuben,	Hornby,	1814	1824	Levi Nash.
Steuben,	Hornellsville,	1792	1795	George Hornell.
Steuben,	Howard,	1806	1810	Henry Kennedy.
Steuben,	Jasper,	1807	1811	Nicholas Prutsman.
Steuben,	Lindley,	1790	1793	Eleazer Lindley.
Steuben,	Pultney,	1802	1810	Melchoir Wagener.
Steuben,	Rathbone,	1794	1812	Isaac Tracy.
Steuben,	Thurston,	1813	1814	Paris Wheelock.
Steuben,	Tuscarora,	1801	1806	William Wombaugh.
Steuben,	Urbana,	1793	1795	John Shether.
Steuben,	Wayland,	1806	1815	Benjamin Perkins.
Steuben,	West Union,	1822	1849	John Wiley.
Steuben,	Wheeler,	1799	1802	Silas Wheeler.
Steuben,	Woodhull,	1805	1806	Caleb Smith.
Suffolk,	Huntington,	1653	1688	Jonathan Rogers.*
Suffolk,	Riverhead,	1690	1659	John Tucker.†
Suffolk,	Smithtown,	1650	1789	George Phillips.
Suffolk,	Southold,	1640	1659	John Tucker.
Sullivan,	Bethel,	1798	1805	John K. Beeman.
Sullivan,	Callicoon,	1814	- -	Jacob Quick.
Sullivan,	Fallsburgh,‡	1788	1795	William A. Thompson.
Sullivan,	Forestburgh,	1793	1783	Capt. A. Cuddeback.

* Rogers was granted permission to build a sawmill at Cold Spring on condition that he furnish lumber at a certain price "and deliver up the stream when the town wants it for a gristmill."

† In 1659 John Tucker "propounded for liberty to sett up a sawmill neere the head of the river, and liberty to cut all sorts of timber, but noe more of oak than fell in the common track of getting pine and cedar which were the chief inducements of getting a mill there to saw."

‡ Charles Webb, who made a survey of the Minnisink Patent in 1762, makes mention in his field notes of Reeves's sawmill.

BEGINNINGS OF THE LUMBER INDUSTRY — CONTINUED.

COUNTY.	TOWN	FIRST SETTLEMENT.	FIRST SAWMILL.	BUILT BY.
Sullivan,	Fremont,	1780	1792	Aaron Pierce.
Sullivan,	Highland,	1825	- -	N. Patterson.
Sullivan,	Liberty,*	1790	1799	Capt. Chas. Brodhead.
Sullivan,	Mamakating,	1728	1730	Manuel Consalus.
Sullivan,	Thompson,	1749	1795	William A. Thompson.
Sullivan,	Tusten,	1763	1760	John Moore.
Tioga,	Barton,	1787	1803	George W. Buttson.
Tioga,	Berkshire,	1791	1810	Bull & Brown.
Tioga,	Candor,	1785	1829	Orange F. Booth.
Tioga,	Newark Valley,	1791	1830†	Patterson & Day.
Tioga,	Nichols,	1787	1833	George Kirby.
Tioga,	Owego,	1785	1826	Willard Foster.
Tioga,	Richford,	1809	1818	Caleb Arnold.
Tioga,	Spencer,	1794	1800	Edmund Hobart.
Tioga,	Tioga,	1789	1792	Major Wm. Ransom.
Tompkins,	Caroline,	1795	1800	Gen. John Cantine.
Tompkins,	Danby,	1795	1797	Dumond & Yaple.
Tompkins,	Dryden,	1797	1800	Ruluff Whiting.
Tompkins,	Enfield,	1804	1812	Benjamin Ferris.
Tompkins,	Groton,	1798	1811	Jonas Williams.
Tompkins,	Ithaca,	1789	1813	Phineas Bennett.
Tompkins,	Newfield,	1801	1809	Eliakim Dean.
Tompkins,	Ulysses,	1791	1796	David Atwater.
Ulster,	Denning,	1821	1827	Dewitt & Reynolds.
Ulster,	Gardiner,	1724	1794	James Jenkins.
Ulster,	Hardenburgh,	1800	1860	Hiram Seager.
Ulster,	Kingston,	1650	1847	Peter J. DuBois.
Ulster,	Marlborough,	1772	1780	James Hallock.

*This mill, which stood on the outlet of Brodhead Pond, was built wholly of logs and hewed timber. The race was constructed from large hemlock trees, "with much labor and ingenuity."

†Steam mill. There may have been a water mill in the town before this one was built; but if so, there is nothing in the town records to indicate it.

BEGINNINGS OF THE LUMBER INDUSTRY — CONTINUED.

COUNTY.	TOWN.	FIRST SETTLEMENT.	FIRST SAWMILL.	BUILT BY.
Ulster,	Olive,	1740	1765	Lemuel Winchell.
Ulster,	Plattekill,	1780	1800	Andrew Garrison.
Ulster,	Rochester,	1688	1703	——— ———.
Ulster,	Saugerties,	1710	1800	Hendrick Schoonmaker.
Ulster,	Shandaken,	1800	1860	D. C. Dutcher.
Ulster,	Shawangunk,	1680	- -	——— Harris.
Ulster,	Woodstock,	1770	1820	Robert Livingston.
Warren,	Bolton,	1792	1813	William H. Oglesva.
Warren,	Chester,	1789	1790	Jabez Mead.
Warren,	Hague,	1796	1833	John Patchen.
Warren,	Horicon,	1802	1839	William Gardner.
Warren,	Johnsburg,	1784	1790	John Thurman.
Warren,	Luzerne,	1797	1815	Jeremiah Rockwell.
Warren,	Queensbury,	1762	1764	Moses Clements.
Warren,	Stony Creek,	1795	1825	James McDonald.
Warren,	Thurman,	1799	1815	Caleb Reynolds.
Warren,	Warrensburgh,	1784	1794	Jonathan Vowers.
Washington,	Cambridge,	1762	- -	Philip Van Ness.
Washington,	Dresden,	1784	1815	Amos Collins.
Washington,	Easton,	1762	1768	Nathan Tefft.
Washington,	Fort Ann,	1781	1788	Jacob Van Wormer.
Washington,	Fort Edward,	1764	- -	William Duer.
Washington,	Granville,	1772	1787	Nathaniel Spring.
Washington,	Greenwich,	1766	1773	Daniel Rose.
Washington,	Hebron,	1770	- -	William Lytle.
Washington,	Kingsbury,	1764	1768	Albert Baker.
Washington,	Putnam,	1784	1802	Robert Cummings.
Washington,	White Creek,	1765	1790	Hercules Rice.
Washington,	Whitehall,	1761	1766	Col. Philip Skene.
Wayne,	Arcadia,	1791	1801	Joseph Caldwell.
Wayne,	Butler,	1803	1819	Jacob S. Viele.
Wayne,	Galen,	1800	1810	Thomas Beadle.

J. Y. MC CLINTOCK, PHOTO.

MODERN REFUSE BURNER.

FOR DISPOSING OF SAWDUST, SLABS AND OTHER WASTE MATERIAL.

BEGINNINGS OF THE LUMBER INDUSTRY — CONCLUDED.

COUNTY.	TOWN.	FIRST SETTLEMENT.	FIRST SAWMILL.	BUILT BY.
Wayne,	Huron,	1796	1809	Elihu Spencer.
Wayne,	Lyons,	1789	1800	John Perrine.
Wayne,	Ontario,	1806	1811	Freeman Hopkins.
Wayne,	Palmyra,	1790	1795	Joel Foster.
Wayne,	Rose,	1805	1811	Elijah Howe.
Wayne,	Savannah,*	1808	1824	Royal Torrey.
Wayne,	Sodus,	1795	1800	Captain Williamson.
Wayne,	Williamson,	1803	1806	Jeremiah Selby.
Wayne,	Wolcott,	1807	1813	Elisha Plank.
Westchester,	Scarsdale,	1701	1668	William Saxton.
Wyoming,	Arcade,	1808	1811	Maj. Moses Smith.
Wyoming,	Bennington,	1802	1808	Chauncey Loomis.
Wyoming,	Castile,	1808	1811	Robert Whalley.
Wyoming,	Covington,	1807	1812	Sprague & Spaulding.
Wyoming,	Eagle,	1808	1811	Amos Huntley.
Wyoming,	Gainesville,	1805	1809	Wheelock Wood.
Wyoming,	Genesee Falls,	1804	1815	Mumford, Smith & McKay.
Wyoming,	Middlebury,	1802	1809	A. Worden.
Wyoming,	Orangeville,	1805	1810	Robert Hopkins.
Wyoming,	Pike,	1806	1809	Eli Griffith.
Wyoming,	Sheldon,	1804	1806	William Vary.
Wyoming,	Wethersfield,	1810	1812	Calvin Clifford.
Yates,	Barrington,	1800	1806	William Cummins.
Yates,	Benton,	1789	1790	Dr. Caleb Benton.
Yates,	Italy,	1793	1795	Asahel Stone, Jr.
Yates,	Jerusalem,	1789	1795	Daniel Brown.
Yates,	Middlesex,	1789	- -	John Walford, Jr.
Yates,	Potter,	1788	1792	Arnold Potter.
Yates,	Starkey,	1800	1807	Timothy Hurd.

* The stream, Crusol Creek, upon which the first sawmill was built, diminished, as the forests about its sources disappeared, so that in about twenty years the mill lost its power. There is now no water power afforded by any stream in the town.